Marketing the Water-Energy-Food (WEF) Nexus

Q. J. Yao

Marketing the Water-Energy-Food (WEF) Nexus

Testing Message Effects with American College Students

palgrave
macmillan

Q. J. Yao
Lamar University
Beaumont, TX, USA

ISBN 978-3-031-93147-5 ISBN 978-3-031-93148-2 (eBook)
https://doi.org/10.1007/978-3-031-93148-2

This Palgrave Macmillan imprint is published by the registered company Springer Nature Switzerland AG.
The registered company address is: Gewerbestrasse 11, 6330 Cham, Switzerland

Preface

Public awareness and support are vital but undervalued for scientific advancement, because public sentiment and opinion largely determine funding and social morale for scientific research. It has been challenging, however, to communicate to the public about scientific matters, because such matters usually require knowledge preparation to understand. How can strategic science and environmental communicators effectively promote scientific matters to the public to garner the necessary support for advancement, particularly using message construction techniques? How does the audience mentally process and respond to such persuasion with different message characteristics? These key questions have intrigued me for a long time and are the focus of this book.

Although the specific topic of this book, the water-energy-food (WEF) nexus, is only about a decade old at the time of this writing, this book reflects a longer period of my work. In 2007, after publishing a study on the development of Chinese environmentalists, I decided to conduct my dissertation project on the influence of environmental discourses among U.S. presidents, the news media, and the American public, under the supervision of Professor Lowndes F. Stephens. The five-level model of communication effects (attention, agenda, knowledge, attitude, and behavior) tested in this book on the WEF nexus was first proposed in that dissertation, which was successfully defended a year later. Since then, while teaching advertising and other communication courses at various universities, I have conducted and published studies about the effects of message factors, mainly sidedness and conclusiveness, on environmental issues such as climate change and, more recently, the WEF nexus.

In 2016, when I presided over the roughly three-year-long and two-round National College Student Science Literacy Survey (NCSSLS), on what is probably the most representative sample of American college students available so far, I incorporated a single-message stimulus into the second round of the survey to test popular message factors (sidedness, conclusiveness, and gain- or loss-framing) and the five-level model of message effects on the WEF nexus, ultimately resulting in this book. The newness of the topic of the WEF nexus carries few social, cultural, or political predispositions and provides an ideal context to observe the persuasion effects. The audience, American college students, although not demographically very similar to the general public, signals the future and hope for scientific research.

Along the way, many people have guided, helped, or collaborated; I express sincere gratitude to them. I am deeply grateful for the guidance of Dr. Stephens, without whom my academic skills could not have been developed, the support of Dr. Jerry Lin and Mr. R. Gavin Jones, for the introduction of the WEF-nexus topic and the funding for the NCSSLS project from the Lamar University (LU) Center for Advances in Water and Air Quality under their leadership, and the collaboration of Dr. Chiung-Fang Chang and Dr. Praphul Joshi on the NCSSLS project and some following publications. I also thank several former LU undergraduate and graduate students, Chelsea McDonald, Jason Tran, Shiyue Hou, and Willam Wheeler, for assisting in the NCSSLS project and some related studies. Special appreciation also goes to anonymous reviewers and other experts who have helped improve this book, particularly to Palgrave Macmillan editor Dr. Robin James, production editor Mr. Heggidi Raju, and project manager Venkitesan Mr. VinodhKumar, who oversaw the book's finalization and publication. Errors, of course, are my responsibility.

Most importantly, I am in debt to my family in the U.S., China, and other places, especially my wife Josie and kids Elizabeth and Jason, for their love and enduring support that make all I have accomplished fulfillable.

Beaumont, TX Q. J. Yao

Competing Interests

The author has no competing interests to declare that are relevant to the content of this manuscript.

Contents

About the Author

Q. J. Yao Professor of Communication and Media at Lamar University (Beaumont, TX), obtained his master's degree in Logic from Beijing Normal University (Beijing, China) and Ph.D. in Mass Communication from the University of South Carolina (Columbia, SC). He has taught strategic communication and other courses at the University of Iowa, Fort Hays State University (Hays, KS), and LU. His research interests lie in examining the persuasive effects of mass, digital, and social media, publishing in such journals as the *Asian Journal of Communication*, *China Advertising*, *Environment Systems and Decisions*, *Environment, Development and Sustainability*, *European Journal of East Asia Studies*, *European Politics and Society*, *Integrative Psychological and Behavioral Science*, the *Journal of International Crisis and Risk Communication Research*, *Journal of Internet Law*, *Journal of Marketing for Higher Education*, *Journal of Media and Religion*, *Journal of Research in Interactive Marketing*, *Public Relations Review*, *Science Communication*, and *Telematics and Informatics*, and other venues.

List of Figures

List of Tables

CHAPTER 1

The Need for Environmental and WEF-Nexus Promotion

In 1962, a milestone book, now globally well known, was published and almost immediately raised significant public attention and discussion about the harm that pesticides and the chemical industry had posed to the environment as well as policymakers' unquestioning acceptance of the promotional information from the chemical industry. If no solutions were proposed and executed, the book warns, all insects and birds could be killed, leaving humans with a "silent spring," as the book is titled, in addition to other health and safety hazards people would be exposed to (Carson, 1962). This dire picture of losing a season that people have cherished and praised in literature for thousands of years frightened the public worldwide. Popular movements were launched, resulting in the establishment of the U.S. Environmental Protection Agency (EPA), much stricter regulation of pesticides and the chemical industry in the U.S., and the era of environmental communication (Epstein, 2014; Yao, 2008). While some scholars believe that the concept of environmental communication was conceived as early as the late nineteenth century when John Muir and Gifford Pinchot addressed environmental preservation and sustainable forestry (Norton, 1995; Smith, 1998), the publication of *Silent Spring*, with the splash it made, was no doubt its first heyday (Epstein, 2014). The book has stimulated a crucial and institutional shift in agricultural policy structure and, after 60 years of publication, remains inspiring and impactful, although the challenge it presents is still daunting, as the voice for

Q. J. Yao, *Marketing the Water-Energy-Food (WEF) Nexus*,
https://doi.org/10.1007/978-3-031-93148-2_1

agri-chemical usage has somehow increased in the past decades (Gareau et al., 2022).

Environmental communication has undergone fundamental transformation since its inception. Initially focusing on environmental stewardship, conservation, and pesticide overuse, environmental communication has gradually expanded to encompass more pressing issues related to human existence, such as air and water pollution, climate change, and the shortage and crises of water, energy, and food (Hoff, 2011; United Nations, 2014; UN News, 2022, August 2; World Economic Forum, 2011). It has shed the misperception of being merely discussions of aesthetic appreciation or bourgeoisie taste and gained recognition as addressing crucial issues that matter to the sustainable livelihood of the planet's residents. The globe is warming at its fastest speed toward "an unacceptable risk," hundreds of millions of people have no safe water, energy, and food, and millions of species are declining or becoming extinct (Gadgil et al., 2022). Scientists and governments from around the world have collaborated to solve those environmental crises for half a century since the rise of the environmental popular movements. However, progress is now best described as "one step forward and two back," due to human beings' "individualistic, materialistic, explorative short-term thinking," economic-growth focus, and political and social conflicts, which are sometimes intense, such as on the topic of climate change (Ivanova & Lele, 2022; Yao et al., 2024). More effective strategic environmental communication, a type of social marketing (Evans et al., 2014; McKenzie-Mohr et al., 2012), is urgently demanded to mobilize collective and coordinated global actions to address these environmental crises (Gadgil et al., 2022) and the transition to a greener and more sustainable lifestyle and economy (Endo et al., 2017).

Among the topics in environmental communications, the water-energy-food (WEF) nexus is a relative newcomer that emerged in 2011 and has already been deemed by leading scientists and international organizations as an effective solution to the water, energy, and food shortages and crises that the world faces and will continuously be challenged with (Hoff, 2011; International Institute for Sustainable Development, 2013; United Nations, 2013a; United Nations, 2013b; United Nations, 2014; World Economic Forum, 2011). The holistic, integrative, and systematic approach of the WEF nexus addresses the UN's Sustainable Development Goals (SDGs), particularly Goals 2, 6, 7, and 12, which pertain to clean water, zero hunger, sanitation, affordable clean energy, and responsible consumption and production. This nexus concept, however, has not been widely adopted by policymakers, incorporated into policy mechanisms, and recognized by the public (Ali & Acquaye, 2024). How can this concept be strategically and effectively communicated to raise public awareness and support, avoid unnecessary politicization, and garner sufficient public resources to address water, energy, and food issues before the situation drastically deteriorates? How can some commonly used message factors affect the persuasiveness of the WEF-nexus strategic communications, and how may those communications be processed by the audience? This book, very likely the first one on this topic, aims to provide answers to these critical questions.

This book is thus intended for readers who are interested in (1) understanding the persuasiveness of popular message factors and audience responses at different levels; (2) communicating critical environmental issues, particularly the WEF nexus, to the public for the best social-marketing effects; and (3) learning more about the attitudes toward science and the environment of American college students. As the subjects of the studies in this book, their characteristics are captured by the representative sample of the first round of the National College Student Science Literacy Survey (NCSSLS) data.

The book consists of six more chapters after this chapter: Chap. 2 discusses the topic of the studies of persuasion effects in the book—the WEF nexus; Chap. 3 further discusses strategic environmental communication as persuasion and social marketing; Chap. 4 describes the NCSSLS survey, whose second-round data will be used to test the hypotheses about the persuasiveness of the message factors and levels in the following chapters; Chap. 5 reviews studies about message levels (agenda/salience, knowledge/information, attitude, and behavior), proposes the hypotheses about the persuasiveness of these message levels, and tests these hypotheses with the survey data; Chap. 6 reviews studies about the message factors (sidedness, conclusiveness, and gain- or loss-frame) and tests the message factors and their interactions; Chap. 7 provides an overall conclusion and discussion of the findings of the studies. The Appendix includes additional information and materials about the NCSSLS survey.

References

Ali, S. M., & Acquaye, A. (2024). An examination of water-energy-food nexus: From theory to application. *Renewable and Sustainable Energy Reviews, 202*, 114669. https://doi.org/10.1016/j.rser.2024.114669

Carson, R. (1962). *Silent Spring*. Houghton Mifflin Company.

Endo, A., Tsurita, I., Burnett, K., & Orencio, P. M. (2017). A review of the current state of research on the water, energy, and food nexus. *Journal of Hydrology: Regional Studies, 11*, 20–30. https://doi.org/10.1016/j.ejrh.2015.11.010

Epstein, L. (2014). Fifty years since Silent Spring. *Annual Review of Phytopathology, 52*, 377–402. https://doi.org/10.1146/annurev-phyto-102313-045900

Evans, W. D., Pattanayak, S. K., Young, S., Buszin, J., Rai, S., & Wallace Bihm, J. (2014). Social marketing of water and sanitation products: A systematic review of peer-reviewed literature. *Social Science & Medicine, 110*, 18–25. https://doi.org/10.1016/j.socscimed.2014.03.011

Gadgil, A., Tomich, T. P., Agrawal, A., Allouche, J., Azevedo, I. M. L., Bakarr, M. I., Jannuzzi, G., Liverman, D., Malhi, Y., Polasky, S., Roy, J., Ürge-Vorsatz, D., & Wang, Y. (2022). The great intergenerational robbery: A call for concerted action against environmental crises. *Annual Review of Environment and Resources, 47*(1), 1–4. https://doi.org/10.1146/annurev-environ-061322-013248

Gareau, B. J., Huang, X., Gareau, T. P., & Didonato, S. (2022). *Silent Spring* at 60: Assessing environmentalism in the cranberry treadmill of production in Massachusetts. *Journal of Rural Studies, 95*(7), 505–520. https://doi.org/10.1016/j.jrurstud.2022.10.001

Hoff, H. (2011, November 16–18). Understanding the nexus: Background paper for the Bonn 2011 Nexus Conference. *Stockholm Environment Institute Website*. https://uploads.water-energy-food.org/resources/SEI-Paper-Hoff-UnderstandingTheNexus-2011.pdf

International Institute for Sustainable Development. (2013). The water–energy–food security nexus: Towards a practical planning and decision-support framework for landscape investment and risk management. *International Institute for Sustainable Development Website*. Retrieved March 4, 2025, from https://www.iisd.org/system/files/publications/wef_nexus_2013.pdf

Ivanova, M., & Lele, S. (2022). Fifty years after UN environment summit, researchers renew call for action. *Nature, 606*, 30. https://doi.org/10.1038/d41586-022-01511-7

McKenzie-Mohr, D., Lee, N. R., Schultz, W. P., & Kotler, P. (2012). *Social marketing to protect the environment: What works*. Sage Publications.

Norton, B. G. (1995). Moralists and aggregators: The case of Muir and Pinchot. In B. G. Norton (Ed.), *Toward unity among environmentalists* (pp. 17–38). Oxford University Press. https://doi.org/10.1093/oso/9780195093971.003.0007

Smith, M. B. (1998). The value of a tree: Public debates of John Muir and Gifford Pinchot. *The Historian, 60*(4), 757–778. https://www.jstor.org/stable/24452183

UN News. (2022, August 2). American West faces water and power shortages due to climate crisis: UN environment agency. *United Nations Website*. Retrieved March 4, 2025, from https://news.un.org/en/story/2022/08/1123782

United Nations. (2013a). *An innovative accounting framework for the food-energy-water nexus: Application for the MuSIASEM approach to three case studies*. Food and Agriculture Organization of the United Nations. Retrieved March 4, 2025, from http://www.fao.org/docrep/019/i3468e/i3468e.pdf

United Nations. (2013b). The status of the water–food–energy nexus in Asia and the Pacific. *UNESCAP.org*. Retrieved March 4, 2025, from http://www.unescap.org/sites/default/files/Water-Food-Nexus%20Report.pdf

United Nations. (2014). The water-energy-food nexus: A new approach in support of food security and sustainable agriculture. *FAO.org*. Retrieved March 4, 2025, from http://www.fao.org/3/a-bl496e.pdf

World Economic Forum. (2011). *Water security: The water–energy–food–climate nexus*. Island Press. http://www3.weforum.org/docs/WEF_WI_WaterSecurity_WaterFoodEnergyClimateNexus_2011.pdf

Yao, Q. (2008). Exploring the social dynamics in the U.S. democracy: Presidential and public opinions about, and media coverage of, environmental issues. Doctoral dissertation, University of South Carolina.

Yao, Q., Chang, C.-F., Joshi, P., & McDonald, C. (2024). Climate change versus the water-energy-food nexus: The oldness or newness of the scientific issue as a factor in the deficit model and the hierarchy of response model. *Environment, Development and Sustainability, 26*, 823–840. https://doi.org/10.1007/s10668-022-02735-3

CHAPTER 2

The WEF Nexus

Before proceeding to test the persuasiveness of the message factors and message levels on the topic of the water-energy-food (WEF) nexus, the significance of this topic needs to be examined first.

People's concern for the planet we inhabit has reached new heights: albeit its deniers remain influential, climate change has received significant global attention (Kimball, 2022; Vázquez et al., 2021) and has been embedded in the United Nation's (UN) Sustainable Development Goals (SDGs; Ali & Acquaye, 2024). In preparing for the 2023 UN Climate Change Conference, commonly known as COP 28, held in Dubai, the UN climate chief warned that everyone is on the frontline of climate and environmental disasters and that world leaders must act to reduce carbon emissions immediately (Harvey, 2023). Less known are other environmental issues that are also vital for our earth's thriving, such as the water-energy-food (WEF) nexus, which are also addressed in the UN SDGs and substantially related to climate issues (Herrera-Franco et al., 2023). Research on effective communication of these issues is increasingly urgent.

1 Our Earth and Its Natural Resources

In literature, Earth or the natural ecosystem it maintains for human living is often personified as "mother," as in "Mother Earth" or "Mother Nature" (Liu et al., 2019; Sacchi et al., 2013). It has benefited all living

Q. J. Yao, *Marketing the Water-Energy-Food (WEF) Nexus*,
https://doi.org/10.1007/978-3-031-93148-2_2

species with its unique resources, some of which are essential to human survival (e.g., oxygen, water, the appropriate atmospheric temperature), and an ecosystem that no other planet has. Humanity's appreciation for "Mother Earth" or "Mother Nature" is probably immaculately expressed in Henry David Thoreau's poem *Nature*:

> O Nature! I do not aspire
> To be the highest in thy quire,—
> To be a meteor in the sky,
> Or comet that may range on high;
> Only a zephyr that may blow
> Among the reeds by the river low;
> Give me thy most privy place
> Where to run my airy race.
> In some withdrawn, unpublic mead
> Let me sigh upon a reed,
> Or in the woods, with leafy din,
> Whisper the still evening in:
> Some still work give me to do,—
> Only—be it near to you!
> For I'd rather be thy child
> And pupil, in the forest wild,
> Than be the king of men elsewhere,
> And most sovereign slave of care:
> To have one moment of thy dawn,
> Than share the city's year forlorn.

As much as human beings appreciate their "Mother Nature," concerned by the increasing deterioration of the natural environment, due to excessive exploitation and neglectful maintenance, and the terrifying future, forward-looking people have been searching for another possible home in the universe for human beings. Although some scientists think that Mars can be transformed into a planet inhabitable for human beings, no one is sure that such transformation is fulfillable in the lifetime of those living on Earth now or even several generations afterward (Wordsworth, 2019). Partly for scientific exploration, partly to address the public curiosity about finding a substitute for an uninhabitable Earth in the future, NASA, the leading institution of astronomy research in the world, has searched for another watery planet, or "another Earth" (NASA, 2015). Its most promising discovery to date is an exoplanet located 300 light-years away (Rice, 2020), not something that humans can reach in an imaginable amount of time.

Thus, the best option is to identify the most efficient approach to consuming Earth's natural resources sustainably and persuade everyone to adopt it. By doing so, Earth will be kept for humans' living rather than continuously depleted toward uninhabitability.

2 Reasons for the WEF Nexus: Issues with Separate Resource Governing

Among all the natural resources that "Mother Earth" provides, the most notable are water, energy, and food, or the ingredients of food. Those natural resources, particularly water, were once taken for granted and inexhaustible. Today, scientists have more than ever realized that effective and sustainable governance of water, energy, and food resources is interwoven with human social and economic developments and that the interplay is intensifying with the growing scarcity of natural resources and climate change (Salam et al., 2017). Such effective governance to ensure responsible and long-lasting usage of those vital natural resources, however, has not been established due to the lack of understanding of the trade-offs among the consumption of those resources (Ali & Acquaye, 2024; Stillwell, 2015), the low public endorsement of such research (Scott et al., 2015; Yao et al., 2018), and the insufficient funds for it (Bhaduri et al., 2015).

So, the dominating governance of natural resources consumption still focuses only on solving crises of a single resource, ignoring the detrimental effects that those crises or their solutions bring to other resources and the deteriorating conflict between the constraint of resources and the expansion of the human population, urbanization, and globalization (Salam et al., 2017; Yao et al., 2018). For instance, although policymakers may have foreseen that global agricultural production will have to increase by approximately 70% by 2050 and that energy production needs to increase by about 50% by 2035 to meet human needs, they do not foresee the impacts of those increases on water, land, and climate resources (Hoff, 2011; United Nations, 2014). Managing those resources separately has worsened water security, the volatility in food prices, and therefrom the global economic system (World Economic Forum, 2011). Here are some cases: Singapore, an advanced civil society with long-term water issues, has to develop technologies to recycle and filter sewage water for drinking purposes (BBC, 2022b, May 25). China, in 2014, after decades of debating and preparation, completed a $56 billion canal-tunnel project, perhaps the largest of its type in the world, to divert 44.8 billion cubic waters from south-middle China to its north area to solve the large-scale issue of water scarcity (Kaiman, 2014). Many places in the world are in a situation close to water bankruptcy (Mabhaudhi et al., 2022; World Economic Forum, 2011). Even in the U.S., the demand for water for daily lives, agricultural irrigation, and industrial production steadily increases, while water supply, mainly from the rivers' flows, has shrunk by about 20% in the past century. For a stunning example, Lake Mead, the largest reservoir in the U.S., has lost 65% of its water storage since 2000 (Kann et al., 2021), threatening the water use for drinking, washing, irrigating, and electricity production for millions of people living in Arizona, California, Colorado, Nevada, New Mexico, and Wyoming, and even Mexico (UN News, 2022, August 2).

The traditionally separated approach to governing global natural resources also disturbingly contributes to global inequality (Mabhaudhi et al., 2022). While food has been stunningly wasted and natural resources are over-exploited in some portions of the world, approximately one-seventh of the global population—called the "bottom billion"—has limited access to fresh water and food supplies that are sufficient for daily living. The rapid increase in the global population is only worsening the water and food shortage and global inequality, and the volume of urban slum dwellers is projected to climb from one billion now to two billion by 2030 (Hoff, 2011). Large-scale coordinating projects like China's South-to-North Water Diversion project (Kaiman, 2014) can be done within a country with the presiding of a national government but is not feasible to solve resource crises on a global scale, due to the lack of international organizations that are similar to national governments in nature, although in the more integrated and collaborative regions of the world like Europe, such international coordinating agencies as the Danube Commission, the International Commission for the Protection of the Danube River (ICPDR), and the International Commission for the Protection of the Rhine (ICPR) have been instrumental in preserving ecosystems across countries (Miao & Nduneseokwu, 2024).

Scientists have recently realized that water, energy, and food are intricately linked in a nexus and should be analyzed and managed in a nexus approach (Brouwer, 2022). The world's agriculture, the primary system for preparing food ingredients, for example, consumes about 70% of global freshwater. Global food production and transportation, on the other hand, account for about 30% of global energy consumption. The world's energy generation also consumes a large share of water. The new technology of hydraulic fracturing, or fracking, for instance, is controversial and largely abandoned in many countries because its operation consumes huge amounts of water, and it can degrade the quality of groundwater as well as cause earth tremors (BBC, 2022a, October 26). A nexus approach to managing natural resources worldwide, leading scientists and international organizations have started to believe, can improve the security of natural resources and ensure more coherent consumption of them globally (United Nations, 2014).

3 Emergence of the Concept of the WEF Nexus

The WEF nexus is still a fairly new scientific concept to most scientists and, particularly, the public (Brouwer, 2022; Yao et al., 2018), which promotes an integrated approach to coordinating the management of the three natural resources simultaneously, balancing the trade-offs for different or competing global interests, and achieving a policy goal for maintaining a natural resource for one sector (e.g., mitigating water shortage) without unnecessarily compromising resources for other sectors (Purwanto et al., 2021). Acknowledging the interconnection and interdependence among water, energy, and food, the WEF nexus approach pursues maximizing mutually beneficial outcomes and minimizing side effects by implementing policies that improve resource-use efficiency and internalize social and environmental impacts in a variety of ranges of spheres and scales (Albrecht et al., 2018).

First emerged in the 2011 Bonn Nexus Conference, the concept of the WEF nexus officially upholds that water, energy, and food are the three pillars of the nexus critical to human survival; they should be analyzed and managed coordinately (Ali & Acquaye, 2024; Endo et al., 2017). The conference's extensive discussions of the concept (Albrecht et al., 2018; Purwanto et al., 2021) are summarized in the background paper, *Understanding the Nexus*, which describes the nexus as "an approach that integrates management and governance" of water, energy, and food "across sectors and scales" as well as regions and production systems (Hoff, 2011). The paper notes that the water issue has been long ignored in the green economy debate, which damages the global ecosystem services, jobs, and livelihood. To adopt the nexus approach and improve water resources and natural infrastructures for global mutual benefits, the paper proposes to: (1) enhance the productivity of resources; (2) recycle and reuse the waste; (3) encourage efficient resource usage through policy incentives; (4) be consistent in resource governance, institutions, and policies; (5) invest in productive ecosystems; (6) balance poverty alleviation and green growth; (7) build capacity and raise awareness; (8) pursue a green economy. Immediately after that conference, a website (https://www.water-energy-food.org/start/) was launched by the German government to serve as an information hub of the nexus concept and facilitate further research. The concept is dynamically supported by the 17 UN SDGs developed in 2015 (Ali & Acquaye, 2024).

The nexus concept soon caught the attention of scientists and leading international organizations that monitor the human ecosystem and the governance of critical natural resources. The World Economic Forum's (2011) Water Initiative published the book *Water Security: The Water–Food–Energy–Climate Nexus*, declaring water as the centerpiece of this nexus that structures those vital natural resources together. The book warns the world's economic leaders that the global population and economic growth are intensifying the demands for water (particularly), energy, and food in the next decades, that those issues are highly interlinked and should be addressed together but the cross-border trading system is failing, and that the poor global collaboration to solve those issues together is threatening the process of globalization. It is the second milestone to establish the WEF-nexus concept after the Hoff (2011) report (Albrecht et al., 2018).

The United Nations has also been playing a leading role in promoting the nexus concept. Its Economic and Social Commission for Asia and the Pacific issued the discussion paper *The Status of the Water–Food–Energy Nexus in Asia and the Pacific* (United Nations, 2013b), analyzing two areas in Asia (Central Asia and Mekong Basin) and concluding that the existing policy framework isolates the water policies and energy policies. The UN Food and Agriculture Organization has also published two reports: *An Innovative Accounting Framework for the Food-Energy-Water Nexus: Application for the MuSIASEM Approach to Three Case Studies* (United Nations, 2013a) analyzes situations in the Republic of Mauritius, The Indian State of Punjab, and South Africa, concluding that holistic approaches to managing natural resources enable simultaneous consideration of different perspectives and issues of sustainability; *The*

Water–Energy–Food Nexus: A New Approach in Support of Food Security and Sustainable Agriculture (United Nations, 2014) discusses the intrinsic complexity intertwined among the resources of water, energy, and food, calling for national level policymakers to weigh the implication of bioenergy, water, land, and electricity upon each other to better coordinate the different, or even competing, global interests in those resources. The UN SDGs, established in 2015, are also vigorously linked to the nexus concept (Ali & Acquaye, 2024).

The International Institute for Sustainable Development (2013) also published a report, which summarizes the key arguments, practices, and lessons learned about applying the WEF nexus approach in natural resource management.

With the promotion of these publications from the leading international organizations, in less than a decade, the concept of the WEF nexus became important to researchers and policymakers involved in sustainability, development, and the green economy (Biggs et al., 2015) and is increasingly discussed in the global community (Purwanto et al., 2021).

Discussions have advanced to the applications of the nexus concept. A systematic review of 245 journal articles and book chapters about the WEF nexus (Albrecht et al., 2018) notes that the nexus approach needs to design more specific, reproducible, qualitative as well as social-scientific methods for nexus assessment, focus more on the linkage and interaction among water, energy, and food (the very issue that demands this new approach), and encourage more cross-discipline collaborations. The review also recommends four key features for developing WEF analytical and assessment methods: innovation (quantify connections, novel analyses, mixed methods that combine the quantitative and qualitative approaches, etc.), the influence of context (multi-levels, local conditions, flexible, etc.), collaboration (cross-sectional or disciplinary, participatory, data-sharing, etc.), and implementation (systematic processes, operational methods, assessable results, etc.).

A more recent review of ten WEF frameworks applied in the past decade has also identified several common principles, such as integrating variables outside the nexus (e.g., the climate); analyzing the nexus in an economic, social, and political context; pursuing a sustainable green economy; relying on interdisciplinary and transdisciplinary knowledge; being solution-oriented and leaning toward global trends with local specifics in mind; and working on capacity building and awareness raising (Purwanto et al., 2021). The review also suggests that the WEF framework builders focus more on resource security, systematically and critically evaluate the performance and complexity of stages in the WEF models, promote innovation and knowledge mobilization, build high-quality databases for research and decision-making, and involve stakeholders in the development of the WEF frameworks (Muthu, 2021).

4 The WEF Nexus and Other Sustainability Issues

The WEF-nexus discussions often involve other sustainability issues, the uppermost of which is climate extremization. The World Economic Forum (2011), for example, believes that the nexus connects water, food, energy, and climate, naming it the water-food-energy-climate nexus. Climate change is largely caused by inefficient energy and land use, which in turn accelerates water and energy scarcity (Hoff, 2011). Man-made climate, as stated by former UN Secretary-General Ban Ki-Moon, has led to water dwindling and scarcity, threatening the world's urban populations (World Economic Forum, 2011). Global warming and governments' inadequate responses to the risk threaten water security mainly by melting the major mountain glaciers, the world's largest freshwater sources. The Himalayan Mountain glaciers, which provide water to more than two billion people in Asia, are estimated as to have shrunk twice as fast in the 1990s than they did in the 1980s and are projected to disappear by about the end of this century. The water and electricity shortages in the American West in the summer of 2022 are also created by the "new very dry normal" resulting from climate extremization (UN News, 2022, August 2). Through constricting water supply and other direct ways, extreme climate events also impose and accelerate pressure on the diminishing of other natural resources, creating grain losses, food shortage, energy scarcity, and other economic risks and crises. If not responded to correctly, those issues could eventually cause a reverse of globalization (World Economic Forum, 2011).

Other issues involved in the WEF-nexus discourses are economic development, population growth, environmental risks and management, global governance inefficiencies, geopolitical conflicts, urbanization, and globalization (Purwanto et al., 2021). The nexus, connecting three resources essential to human survival, is vital in our economic and societal systems.

Despite its vital role, the WEF-nexus concept, though only a decade old, is not without criticism. Disagreeing scholars and policymakers criticize the concept as "old wine in the new bottle," because its operations in the local communities are not much different from the traditional approaches to managing the water, energy, and food resources separately. They also attack the concept as vaguely defined, with the frameworks derived from it having less common in operation (Purwanto et al., 2021). Whereas many criticisms are points to improve or gaps for further research, they also evidence that the WEF nexus as a new concept of resource management demands explication and promotion to earn public support and contribute more to solving the deteriorating resource risks (Muthu, 2021). The nexus approach, therefore, needs to further engage the media to effectively inform the public (Endo et al., 2017), whose influence is indispensable in building a policy framework of WEF nexus (United Nations, 2014).

Meanwhile, the newness of the WEF-nexus concept (Endo et al., 2017) renders it a better candidate to study how the public processes scientific information mentally with little predisposition interfering, compared with the complicated or polarized scientific or environmental concepts or issues, such as climate change (Yao et al., 2024). This book, based on the National College Student Science Literacy Survey (NCSSLS) data, whose second round was embedded with a survey experiment that manipulated message levels (agenda, knowledge, attitude, and behavior) and factors (sidedness, conclusiveness, gain- or loss-frame), identifies effective approaches to communicating the WEF nexus to the college students, and suggestively the general public, in a primitive context with fewer confounders. The book raises readers' awareness and support for the nexus approach, which more efficiently mitigates the increasingly challenging relationships among the three dwindling but vitally needed resources (Muthu, 2021), as called by the World Economic Forum (2011). The book also adds to the scholarship of science, environment, and health communications by measuring the cross-sectional status of college students' knowledge of, and attitude and behavioral intention toward, the WEF nexus and its impacts on livelihood.

References

Albrecht, T. R., Crootof, A., & Scott, C. A. (2018). The water-energy-food nexus: A systematic review of methods for nexus assessment. *Environmental Research Letters, 13*, 043002. https://doi.org/10.1088/1748-9326/aaa9c6

Ali, S. M., & Acquaye, A. (2024). An examination of water-energy-food nexus: From theory to application. *Renewable and Sustainable Energy Reviews, 202*, 114669. https://doi.org/10.1016/j.rser.2024.114669

BBC. (2022a, October 26). What is fracking and why is it controversial? *BBC Website*. Retrieved April 7, 2022, from https://www.bbc.com/news/uk-14432401

BBC. (2022b, May 25). NeWater: Turning urine into beer in Singapore. *BBC Website*. Retrieved 2022, May 26, from https://www.bbc.com/news/av/world-asia-61577976

Bhaduri, A., Ringler, C., Dombrowski, I., Mohtar, R., & Scheumann, W. (2015). Sustainability in the water-energy-food nexus. *Water International, 40*(5-6), 723–732. https://doi.org/10.1080/02508060.2015.1096110

Biggs, E., Bruce, E., Boruff, B., Duncan, J. M. A., Horsley, J., Pauli, N., McNeill, K., Neef, A., Van Ogtrop, F., Curnow, J., Haworth, B., Duce, S., & Imanari, Y. (2015). Sustainable development and the water-energy-food nexus: A perspective on livelihoods. *Environmental Science & Policy, 54*, 389–397. https://doi.org/10.1016/j.envsci.2015.08.002

Brouwer, F. (Ed.). (2022). *Handbook on the water-energy-food nexus*. Edward Elgar Publishing.

Endo, A., Tsurita, I., Burnett, K., & Orencio, P. M. (2017). A review of the current state of research on the water, energy, and food nexus. *Journal of Hydrology: Regional Studies, 11*, 20–30. https://doi.org/10.1016/j.ejrh.2015.11.010

Harvey, F. (2023, November 24). World stands on frontline of disaster at COP 28, says UN climate chief. *The Guardian*. Retrieved March 4, 2025, from https://www.theguardian.com/environment/2023/nov/24/un-top-climate-official-simon-stiell-cop-28-dubai

Herrera-Franco, G., Bolmann, H. A., Lofhagen, J. C. P., Bravo-Montero, L., & Carrion-Mero, P. (2023). Approach on water-energy-food (WEF) nexus and climate change: A tool in decision-making processes. *Environment Development, 46*, 100858. https://doi.org/10.1016/j.envdev.2023.100858

Hoff, H. (2011). Understanding the nexus: Background paper for the Bonn 2011 Nexus Conference. SEI Website, November 16–18, 2011. https://uploads.water-energy-food.org/resources/SEI-Paper-Hoff-UnderstandingTheNexus-2011.pdf

International Institute for Sustainable Development. (2013). The water–energy–food security nexus: Towards a practical planning and decision-support framework for landscape investment and risk management. *International Institute for Sustainable Development Website*. Retrieved March 4, 2025, from https://www.iisd.org/system/files/publications/wef_nexus_2013.pdf

Kaiman, J. (2014, December 12). China's water diversion project starts to flow to Beijing. *The Guardian*. Retrieved March 4, 2025, from https://www.theguardian.com/world/2014/dec/12/china-water-diversion-project-beijing-displaced-farmers

Kann, D., Rigdon, R., & Wolfe, D. (2021, August 21). The Southwest's most important river is drying up. *CNN Website*. Retrieved March 4, 2025, from http://www.cnn.com/interactive/2021/08/us/colorado-river-water-shortage/

Kimball, S. (2022, August 17). CDC Director Walensky to reorganize agency after admitting Covid pandemic responses fell short. *CNBC Website*. Retrieved March 4, 2025, from http://www.cnbc.com/2022/08/17/cdc-admits-covid-response-fell-short-launches-reorganization.html

Liu, T., Geng, L., Ye, L., & Zhou, K. (2019). "Mother Nature" enhances connectedness to nature and pro-environmental behavior. *Journal of Environmental Psychology, 61*, 37–45. https://doi.org/10.1016/j.jenvp.2018.12.003

Mabhaudhi, T., Senzanje, A., Modi, A., Jewitt, G., & Massawe, F. (Eds.). (2022). *Water-energy-food narratives and resource securities: A global south perspective* (1st ed.). Elsevier.

Miao, Q., & Nduneseokwu, C. (2024). *Environmental leadership in a VUCA era: An interdisciplinary handbook*. Springer Singapore. https://doi.org/10.1007/978-981-96-0324-4

Muthu, S. S. (Ed.). (2021). *The water-energy-food nexus: Concepts and assessments*. Springer Singapore. https://doi.org/10.1007/978-981-16-0239-9

NASA. (2015, July 23). Finding another Earth. *NASA Website*. Retrieved March 4, 2025, from https://science.nasa.gov/earth/finding-another-earth/

Purwanto, A., Susnik, J., Suryadi, F. X., & de Fraiture, C. (2021). Water-energy-food nexus: Critical review, practical applications, and prospects for future research. *Sustainability, 13*(4), 1919. https://doi.org/10.3390/su13041919

Rice, D. (2020, April 15). Has NASA discovered another Earth? Perhaps. *USA Today*. Retrieved March 4, 2025, from https://www.usatoday.com/story/news/nation/2020/04/15/has-nasa-discovered-another-earth/5139773002/

Sacchi, S., Riva, P., & Brambilla, M. (2013). When Mother Earth rises up anthropomorphizing nature reduces support for natural disaster victims. *Social Psychology, 44*(4), 271–277. https://doi.org/10.1027/1864-9335/a000112

Salam, P. A., Shrestha, S., Pandey, V. P., & Anal, A. K. (Eds.). (2017). *Water-energy-food nexus: Principles and practices*. Wiley/American Geophysical Union.

Scott, C., Kurian, M., & Wescoat, J. L., Jr. (2015). The water-energy-food nexus: Enhancing adaptive capacity to complex global challenges. In M. Kurian & R. Ardakanian (Eds.), *Governing the nexus: Water, soil, and waste resources considering global change* (pp. 15–38). Springer International Publishing. https://doi.org/10.1007/978-3-319-05747-7_2

Stillwell, A. (2015). Sustainability of public policy: Example from the energy-water nexus. *Journal of Water Resources Planning and Management, 141*(12), A4015001. https://doi.org/10.1061/(ASCE)WR.1943-5452.0000522

UN News. (2022, August 2). American West faces water and power shortages due to climate crisis: UN environment agency. *United Nations Website*. Retrieved March 4, 2025, from https://news.un.org/en/story/2022/08/1123782

United Nations. (2013a). *An innovative accounting framework for the food-energy-water nexus: Application for the MuSIASEM approach to three case studies.* Food and Agriculture Organization of the United Nations. Retrieved March 4, 2025, from http://www.fao.org/docrep/019/i3468e/i3468e.pdf

United Nations. (2013b). The status of the water–food–energy nexus in Asia and the Pacific. *ESCAP Website.* Retrieved March 4, 2025, from http://www.unescap.org/sites/default/files/Water-Food-Nexus%20Report.pdf

United Nations. (2014). The water-energy-food nexus: A new approach in support of food security and sustainable agriculture. *FAO.org.* Retrieved March 4, 2025, from http://www.fao.org/3/a-bl496e.pdf

Vázquez, A., Larzabal-Fernández, A., & Lois, D. (2021). Situational materialism increases climate change scepticism in men compared to women. *Journal of Experimental Social Psychology,* 96, Article 104163. https://doi.org/10.1016/j.jesp.2021.104163

Wordsworth, R. (2019, August 21). Can Mars be made habitable in our life? *Scientific American.* Retrieved March 4, 2025, from https://blogs.scientificamerican.com/observations/can-mars-be-made-habitable-in-our-lifetime/

World Economic Forum. (2011). *Water security: The water–energy–food–climate nexus.* Island Press. http://www3.weforum.org/docs/WEF_WI_WaterSecurity_WaterFoodEnergyClimateNexus_2011.pdf

Yao, Q., Chang, C.-F., Joshi, P., & McDonald, C. (2024). Climate change versus the water-energy-food nexus: The oldness or newness of the scientific issue as a factor in the deficit model and the hierarchy of response model. *Environment, Development and Sustainability, 26*, 823–840. https://doi.org/10.1007/s10668-022-02735-3

Yao, Q., Joshi, P., Chang, C.-F., McDonalds, C., Tran, J., Wheeler, W., & Hou, S. (2018). Advocating a new approach to governing water, energy, and food security: Testing the effects of message inoculation and conclusion explicitness in the case of the WEF nexus. *Journal of International Crisis and Risk Communication Research, 1*(1), 111–133. https://jicrcr.com/index.php/jicrcr/article/view/6/6

CHAPTER 3

Promoting the WEF Nexus as Persuasion and Social Marketing

Promoting the water-energy-food (WEF) nexus essentially belongs to strategic science or environmental communication, a type of communication designed with strategies geared toward a goal and a targeted audience to promote scientific or environmental awareness or movements. Such strategic communication, by default, requires deliberative design based on a sufficient understanding of the audience, its mental route to process information, and its responses to the communication in accordance with the communication's goal, objectives, strategies, and tactics (Besley & Dudo, 2022). The effectiveness of strategic science or environmental communication is critically relied upon when tackling crises, like the COVID-19 pandemic. The challenge to achieve effectiveness, nevertheless, urges science and environmental communication researchers to further understand the mechanisms underlying or undermining that effectiveness, such as the resistance to the COVID vaccine (Bolsen & Palm, 2022) or the indifference to the WEF nexus concept when it needs broader public support. That urge for further understanding of strategic science or environmental communications is the motivation for studies of the persuasiveness of the message factors and levels in this book. It aims to build upon the accumulated literature on environmental and WEF nexus promotion, persuasion, and social marketing.

Q. J. Yao, *Marketing the Water-Energy-Food (WEF) Nexus*,
https://doi.org/10.1007/978-3-031-93148-2_3

1 Environmental Promotion as Strategic Science Communication and Its Challenges

Scientists have long realized that sufficient public support is key to the success of their research, not only because public endorsement helps acquire funds and favorable legislation, but also because it generates positive social morale in the research society (Scheufele, 2013; Yao et al., 2024). On the other hand, public opposition can effectively lead to the demise of a scientific field, like what happened to the nuclear energy sector in the U.S. after the Three Mile Island accident in 1979 (Yao, 2008).

A major challenge for researchers and practitioners of science communication, including environmental communication, is that, although science and environmental research and causes need to involve the public for long-lasting public support and engagement, the public may not have enough knowledge preparation to adequately understand issues in these areas, particularly when such issues are heatedly discussed and entangled with countless social, cultural, or political cues. Some scholars, like the young Jurgen Habermas, maintain that the public should fully and actively participate in all discussions of science and environmental issues; others, like the older Habermas, recognize that science and environmental issues require special expertise to understand and make decisions, which many members of the public may have difficulty to obtain (Yao, 2008; Yao et al., 2020; Yao et al., 2024). In either case, however, effective strategic communications to grow public support is indispensable to advance scientific or environmental research.

Many science communication researchers also realize that, despite the crucial role of public support in scientific advancements, the public has largely been indifferent to science or environmental issues, because those issues are relatively distant from the public's daily lives and knowledge base. Public knowledge of science or environmental issues in the U.S., for instance, has been consistently detected as low by the surveys, despite the extensive government-supported science promotion and education programs (Burns et al., 2003), which clearly indicates a "public deficiency" (Miller, 2001, p. 116). According to a popular measurement of civic scientific literacy that started in the 1950s, no more than a quarter of the U.S. and European adult populations can be considered scientifically literate, having adequate knowledge of science or environmental issues to support judicious decisions. Science communicators believe, however, that if adequate information is provided to the public to remove its deficit in the special knowledge of science and environmental matters, the public will then support more, hence the "deficit model" that has long been popular among science communication researchers, which calls for constant feeding of scientific information to the public (Miller, 2001; Scheufele, 2013; Sturgis & Allum, 2004).

Findings of later studies have complicated the "deficit model": supportive attitudes or behaviors are not always generated when providing public science or environmental knowledge, possibly due to some confounding variables (e.g., Sturgis & Allum, 2004). Opponents attribute the lack of a positive correlation between science knowledge and supportive public attitude to three social or cultural factors: (1) highly debated topics complicated by cultural, social, economic, or political factors can disguise the relationship between knowledge and attitude (Yao, 2008; Yao et al., 2024); (2) the public prioritizes coping risks closely related to cultures, value systems, or worldviews over more objective and distant hazards of scientific or environmental issues (Miller, 2001; Sturgis & Allum, 2004); (3) the public has low trust in scientists or science journalists (Critchley, 2008; Master & Resnik, 2013; Scheufele, 2013). The nature of the scientific issues has been empirically identified as a confounder of the knowledge-attitude relationship. On new issues, such as the WEF nexus, providing more information leads to more supportive attitudes and behavioral intentions, while on hotly debated topics such as climate change, the hypothesized linkages among knowledge, attitude, and behavioral intention are not identifiable. On complex issues, even when provided with more information, people can still hold their original attitudes and behavioral intentions shaped by their cultural, political, or religious leaders (Belch & Belch, 2014; Kollmuss & Agyeman, 2002; Scheufele, 2013). More research is needed to understand the effectiveness of communication strategies on these complicated issues (Yao et al., 2024). The hypothesis that the nature of scientific issues makes a difference in public response (Yao et al., 2024) can help explain this phenomenon: the American public's concern for environmental issues, particularly climate change, rose to its apex around Earth Day of 1990, when people just experienced one of the hottest summers in 1988, and has stayed high; the public support, however, has been lukewarm, making tangible progress along the environmental agenda difficult (Gadgil et al., 2022; Obermiller, 1995).

Other researchers have also found that the way to measure knowledge makes a difference: while science or environmental knowledge measured with respondents' subjective perception (subjective knowledge) does not correlate with support, knowledge measured with objective tests (objective knowledge) reveals a positive correlation and confirms the deficit model (Stoutenborough & Vedlitz, 2014). Most researchers still believe that the lack of a positive relationship between knowledge and attitude due to the inference of the cultural, social, economic, or political factors should not invalidate the deficit model; instead, it reveals the complexity of this relationship (Sturgis & Allum, 2004). On newer scientific or environmental issues that have fewer associations with heated cultural or political debates, such as the WEF nexus, the positive linkage between the public's knowledge and attitude should be identifiable (Yao et al., 2018; Yao et al., 2024).

A limited understanding of effective strategies in environmental communication has resulted in the low efficiency in generating strong public support and action on critical matters. In addition, it elicits the incorrect assumption of social marketing that the audience will respond positively, to simple or complex issues, as long as the audience members are informed about those issues. The three goals of the environmental movements, especially strategic environmental communications—environmental protection, economic health, and equity of social justice (the three Es; Pezzullo & Cox, 2021)—have not really been the priorities of people in the U.S. or globally. With alarming projections of dire, imminent, irreversible environmental losses and crises, such as surpassing a global temperature threshold that cannot be remedied for more than a thousand years, identifying more effective environmental communication strategies is urgently needed for environmental scientists and communicators to garner resources and investment and cultivate public support for their causes of environmental protection and sustainable development (Brulle, 2010).

Among efforts to build tools for effective strategic environmental communications, a theory of environmental identity campaigns stands out. It calls for the application of cognitive science in the creation of more effective environmental communication messages that can appeal to the audience's pre-existing values, cognitive characteristics, and self-identity (supposedly also developed by professional strategic communicators) to generate favorable public opinion and thereby supportive legislation. Despite its criticism for possibly falling into an elite-driven communication model, the approach of environmental identity campaigns has shown effectiveness in generating immediate attitudinal and behavioral changes. Those environmental communication strategies based on cognitive science, of course, also need to take into consideration the larger social context for participation, civil engagements, and social changes (Brulle, 2010). They lead the path to study strategic environmental communication as persuasion and in the big context of social marketing research.

2 Strategic Science and Environmental Communication as Persuasion and Social Marketing

The strategy of "nudging" (Thaler & Sunstein, 2009) partly earned Chicago professor Richard Thaler the 2017 Nobel Prize in Economics. "Nudging" means to motivate members of society to engage in socially desirable practices, such as environmental protection, and avoid undesirable ones, such as littering, using persuasion and campaign strategies. These strategies are designed based on principles of social psychology, particularly people's limited rationality, social preferences, and lack of self-control. The prize highlights social marketing research, a line of study that started in the early 1970s but has now gained huge popularity among governments and organizations, including those missioned to enhance environmental sustainability (McKenzie-Mohr et al., 2012).

2.1 Social Marketing, Strategic Communication, and Persuasion

The term "social marketing" was first coined in a pioneering piece (Kotler & Zaltman, 1971), which defines social marketing as "the design, implementation, and control of programs calculated to influence the acceptability of social ideas and involving considerations of product planning, pricing, communication, distribution, and marketing research," namely, the mix of four Ps (Lee & Kotler, 2015). The product in social marketing is the social behaviors that are promoted to adopt or reject (e.g., adopting the WEF-nexus approach), which is more challenging to design than products in business marketing campaigns. Social marketers may need to design some tangible or buyable products to surround and support the core social behaviors. For the promotion element, social marketers should rely on not only social advertising but also personal selling (sending people to promote the acceptance or rejection of the social behaviors personally), publicity (earning news coverage of the acceptance or rejection of the social behaviors), and sales promotion (designing programs or incentives to attract people to accept or reject the social behaviors). The place element requires social marketers to provide adequate ways to distribute information on accepting or rejecting social behaviors and buying supporting tangible products; social marketers also should develop convenient channels to collect people's responses. The final element, price, reminds social marketers to demand only an appropriate amount of people's money, time, energy, and opportunities for them to accept or reject the social behavior or buy tangible products. Overall, the social marketing perspective suggests that governments or NGOs study their targeted audience, design effective communications, events, or objects, and establish social behaviors that are good for the whole society. It can clearly be used to earn public awareness of, and support for, the WEF-nexus approach (World Economic Forum, 2011).

Researchers have identified four core concepts and five persuasion techniques of social marketing (French & Russell-Bennett, 2015). Social marketing must (1) aim at social behavioral influence; (2) be among citizens, customers, or members of civic societies; (3) provide social offerings (ideas, products, services, experiences); and (4) eventually establish a relationship between policymakers or public organizations and the citizens. The five techniques are: (1) integrated intervention mix (through mass, social, and personal channels), (2) competition analysis and action (research on possible negative factors), (3) systematic planning and evaluation (theory- and research-based design of intervention programs), (4) insight-driven segmentation (based on demographics, psychographics, behavioral patterns, etc.), and (5) co-creation through social markets that involve all parties (citizens, stakeholders, and other civic and commercial institutions). A study, for example, reveals that anti-alcohol social marketing campaigns that target adolescents and are designed purely by experts are not as effective as those campaigns that are co-designed by experts and members of the targeted audience because the latter normally have interactive and engaging components (Rodriguez-Sanchez & Sancho-Esper, 2016). The techniques of social marketing are so critical for governments and public organizations to achieve their goals that the Center for Disease Control and Prevention's (CDC) Division for Cancer Prevention and Control has designed an online tool, MessageWorks, to help strategic communicators develop messages for social marketing campaigns aiming to change health-related behaviors.

Social marketing has attracted much attention from scholars interested in using strategic communications to solve policy or public interest issues (Lee & Kotler, 2015), mostly health-related probably due to the global focus on health. One review study identifies 166 articles about social marketing from 242 peer-reviewed journals, mainly about social issues of nutrition, alcohol prevention, HIV/AIDS, and physical activity. The review recommends using formative research to measure campaign effectiveness and understand the targeted audience's characteristics, attitudinal and behavioral patterns, and preferred communication methods (Truong & Dang, 2016). Another systematic review has found 125 studies about social marketing campaigns on global health issues that were conducted in lower- and middle-income countries from three databases (PubMed, PsycInfo, and ProQuest) published between 1995 and 2013, mostly about HIV in sub-Saharan Africa (Firestone et al., 2017). Searching nine databases, a synthetic-review study has identified 23 studies about social marketing campaigns just on the issue of alcohol drinking from 2000 to 2014 (Kubacki et al., 2015). Another synthetic review of 32 articles on social marketing about healthy drinking water and sanitation has confirmed that those marketing campaigns for social causes or services focus on the four Ps as persuasion campaigns for commercial brands do (Evans et al., 2014). Several journals are titled "social marketing," indicating its growing popularity as an approach for strategic communicators to persuade the public to support social causes or ideas.

Social marketing can be largely applied to strategic environmental communication to use marketing persuasion strategies and techniques for the promotion of desirable environmental and healthy behaviors (e.g., sustainable development, eco-friendly consumption, and low-carbon lifestyle) and change undesirable ones (e.g., wasting water or food, drinking bottled water excessively, using only plastic shopping bags). To change citizens' behaviors and benefit the whole society, these campaigns are usually sponsored by governments or concerned non-governmental organizations (NGOs). For example, one of the well-known social marketing campaigns, "Don't Mess with Texas," is sponsored by the Texas Department of Transportation, which has lasted for three decades and persuaded millions of people, possibly more than 95% of the U.S. adult population, not to litter and keep the highway environment clean.

Social marketing is different from nonprofit marketing, public sector marketing, and cause marketing, which are also mostly conducted by governments and NGOs. The goal of social marketing campaigns is to establish socially desirable behaviors or abandon those socially undesirable ones, and it is not measured in financial terms. In nonprofit or public sector marketing campaigns, whose goals are usually promoting products or services provided by those nonprofit or public entities, however, financial return on investment can still be applied to measure the campaign results. Cause marketing campaigns mainly aim to raise awareness and knowledge of social issues, such as climate change or drunk driving, but they may not focus on changing behaviors. Social marketing like strategic environmental communication starts with raising public awareness or knowledge, but pro-social behaviors are the ultimate objective to achieve. When the world faces looming environmental crises that need to be fixed urgently, financial calculations are not involved in strategic environmental communication campaigns.

2.2 *Social Marketing Campaigns on Science and Environment*

Social marketing campaigns have been widely implemented around the world by science and environmental communicators to persuade the public and achieve societal benefits. For instance, in 1995, the Australian government found in a survey that only 33% of Australian children up to six years old met the recommended standard of immunization. To increase their vaccination rate, the government designed and launched the "Immunize Australia" social marketing campaign in 1997. The campaign included several initiatives, including (1) research programs to target segments of parents who possibly neglected their kids' immunization and identify communication barriers that discouraged parents from taking immunization measures for their kids, (2) communication programs directed to those parents that were designed based on theoretical health communication models, and (3) training programs to enhance the knowledge and communication skills of the immunization service providers. The campaign eventually developed an *Australian Immunization Handbook* and distributed it to more than 60,000 immunization workers. It also educated those workers, parents, and communities through television and magazine advertisements and posters in clinics, hospitals, and post offices, as well as events like Immunization Awareness Days. The campaign increased social awareness, knowledge, and even the recommended standard of immunization in Australia. By June 2001, 91% of Australian children were immunized with the new and higher standard of recommendation (Carroll, 2002).

In another case, to increase male Zimbabweans' uptake of voluntary medical male circumcision (VMMC) service to reduce the high infection rate of HIV and AIDS in the country, partnering with several international organizations (such as UNAIDS and USAID), the nonprofit global health organization Population Service International (PSI) launched a social marketing campaign, named "Smart," in 2011. The campaign targeted two male segments: 15–24 years old, and 25–49 years old. VMMC was branded as not only an HIV prevention measure but also a lifestyle choice. Information on, and female support for, the medical procedure were provided through the campaign to ease the males' concerns and draw their attention to the benefits of VMMC. The campaign was conducted during school holidays through mass media communications and community mobilizations. VMMC services were offered in clinics close to the younger male crowds. The research-based social marketing campaign was very successful. Between 2011 and 2014, the number of males taking the VMMC service quadrupled in the country, and the "Smart" campaign gained awareness and positive behaviors nationwide (French, 2017).

In England, between 2009 and 2014, a "Time to Change" anti-stigma social marketing campaign was launched to increase public awareness, knowledge, and actions to tackle mental health issues in the communities. The campaign, designed with persuasion techniques, used mass media, social media, and social contact events. An online survey of 10,526 participants from the English population was conducted at about the same time to measure the campaign results. The awareness of the campaign rose from zero before it to as high as 81.7% at the end, and people who showed a higher level of awareness had a higher level of knowledge of mental issues and a higher level of supportive attitude toward those who had those issues. The campaign's website, Facebook page, and Twitter account, set up when the campaign started, also quickly gained visitors and followers. By the end of 2014, more than 250,000 people had actively used the campaign's social media channels (Henderson et al., 2020).

Persuasion techniques and social marketing campaigns have been demonstrated as instrumental in social movements and have been globally relied on to achieve desirable outcomes on science and health matters. Strategic environmental communicators and researchers, facing the arduous task of urgently changing the public attitude and behaviors to avoid irreversible environmental damages, need to understand and apply these ideas, principles, and techniques to fulfill their challenging task, particularly when that daunting task is often interfered, distracted, or degraded by other seemingly more imminent man-made global crises, such as wars or other international conflicts.

The principles of social marketing have actually already been formally introduced into the field of strategic environmental communication by Philip Kotler, a founding scholar of social marketing research, and his associates (McKenzie-Mohr et al., 2012), who have built a community-based social marketing (CBSM) model for strategic environmental communicators to foster environmentally friendly behaviors (McKenzie-Mohr, 2011, 2022). The model advises strategic environmental communicators to move beyond traditional information-based campaigns and adopt marketing and persuasion techniques to increase campaign effectiveness. Those techniques include (1) carefully selecting the sustainable behavior as the promotional target through weighing these behaviors' impact, penetration, adoption probability, and so on; (2) identifying the barriers and benefits to practice the selected sustainable behavior through literature review, observations, focus groups, surveys or other methods; (3) designing communication strategies, messages, and behavior-change tools (gaining a commitment, setting behavioral goals, developing community norms, producing reminding prompts, using social diffusion, providing helpful goods and services, providing incentives and conveniences, etc.) to reduce the perception of these barriers and increase the perception of the benefits among the target audience; (4) pilot-testing these strategies and tools with a small segment of the target audience community; and (5) evaluating the impacts and modifying the communication strategies and tools for broader implementation (McKenzie-Mohr, 2011; McKenzie-Mohr et al., 2012). The model emphasizes the importance of designing effective strategic persuasion messages for environmental promotion and has been used to promote sustainable behaviors in some college communities. Its effectiveness, based on interpersonal communication in local communities, boosts confidence in identifying persuasion techniques and tools for message creation in large-scale mediated strategic environmental communications.

3 Persuasion Factors: The Persuader, the Message, the Channel, and the Audience

Strategic environmental communications, as a specific type of social marketing, is fundamentally persuasion. It involves the persuader, the message, the channel, and the audience, which strategic environmental communication researchers and practitioners need to understand, respectively, to create effective campaigns (Armstrong, 2010).

In a synthetic review of the growth of the general environmental communication as a field, Robert Cox and Stephen Depoe (2015) identify seven major research questions that environmental communication studies have attempted to answer: (1) How human agents represent nature/environment; (2) how this representation is communicatively shaped developmentally; (3) what effects different environmental sources (e.g., media) and communicative practices have on audiences; (4) what the relationships are between communication and audiences' values, beliefs, and behaviors; (5) how message creation and dissemination influence audience responses; (6) what the relationship between individualistic identity and personal environment is; (7) what the relationship between local culture and environment is. Those questions are also related to strategic environmental communication or persuasion. The first question examines the persuaders and their understanding of the topic. The second involves the persuaders' practices and the message. The third addresses persuasion channels (e.g., media), and the rest explores how social, cultural, economic, or political characteristics of the audience interfere in the receipt of and responses to the persuasion.

Two of these persuasion factors, the audience and the message (i.e., levels of agenda, knowledge, attitude, and behavior; and factors of sidedness, conclusiveness, and gain- or loss-framing), will be the focus of the research presented in the following chapters and analyzed with the NCSSLS data.

References

Armstrong, S. (2010). *Persuasive advertising: Evidence-based principles.* Palgrave Macmillan.

Belch, G. E., & Belch, M. A. (2014). *Advertising and promotion: An integrated marketing communications perspective* (10th ed.). McGraw-Hill/Irwin.

Besley, J., & Dudo, A. (2022). *Strategic science communication: A guide to setting the right objectives for more effective public engagement.* Johns Hopkins University Press.

Bolsen, T., & Palm, R. (2022). Politicization and COVID-19 vaccine resistance in the U.S. *Progress in Molecular Biology and Translational Science, 188*(1), 81–100. https://doi.org/10.1016/bs.pmbts.2021.10.002

Brulle, R. J. (2010). From environmental campaigns to advancing the public dialog: Environmental communication for civic engagement. *Environmental Communication, 4*(1), 82–98. https://doi.org/10.1080/17524030903522397

Burns, T. W., O'Connor, D. J., & Stocklmayer, S. M. (2003). Science communication: A contemporary definition. *Public Understanding of Science, 12*(2), 183–202. https://doi.org/10.1177/09636625030122004

Carroll, T. (2002). Public health social marketing: The Immunize Australia program. *Social Marketing Quarterly, 8*(1), 55–61. https://doi.org/10.1080/15245000212542

Cox, R., & Depoe, S. (2015). Emergence and growth of the 'field' of environmental communication. In A. Hansen & R. Cox (Eds.), *The Routledge handbook of environment and communication (Chapter 1)*. Routledge. https://doi.org/10.4324/9781315887586

Critchley, C. R. (2008). Public opinion and trust in scientists: The role of the research context, and the perceived motivation of stem cell researchers. *Public Understanding of Science, 17*(3), 309–327. https://doi.org/10.1177/0963662506070162

Evans, W. D., Pattanayak, S. K., Young, S., Buszin, J., Rai, S., & Wallace Bihm, J. (2014). Social marketing of water and sanitation products: A systematic review of peer-reviewed literature. *Social Science & Medicine, 110*, 18–25.

Firestone, R., Rowe, C. J., Modi, S. N., & Sievers, D. (2017). The effectiveness of social marketing in global health: A systematic review. *Health Policy Planning, 32*(1), 110–124. https://doi.org/10.1093/heapol/czw088

French, J. (2017). *Social marketing and public health: Theory and practice*. Oxford University Press.

French, J., & Russell-Bennett, R. (2015). A hierarchical model of social marketing. *Journal of Social Marketing, 5*(2), 139–159. https://doi.org/10.1108/JSOCM-06-2014-0042

Gadgil, A., Tomich, T. P., Agrawal, A., Allouche, J., Azevedo, I. M. L., Bakarr, M. I., Jannuzzi, G., Liverman, D., Malhi, Y., Polasky, S., Roy, J., Ürge-Vorsatz, D., & Wang, Y. (2022). The great intergenerational robbery: A call for concerted action against environmental crises. *Annual Review of Environment and Resources, 47*(1), 1–4. https://doi.org/10.1146/annurev-environ-061322-013248

Henderson, C., Potts, L., & Robinson, E. (2020). Mental illness stigma after a decade of Time to Change England: Inequalities as targets for further improvement. *European Journal of Public Health, 30*(3), 497–503. https://doi.org/10.1093/eurpub/ckaa013

Kollmuss, A., & Agyeman, J. (2002). Mind the gap: Why do people act environmentally and what are the barriers to pro-environmental behavior? *Environmental Educational Research, 8*(3), 239–260. https://doi.org/10.1080/13504620220145401

Kotler, P., & Zaltman, G. (1971). Social marketing: An approach to planned social change. *Journal of Marketing, 35*(3), 3–12. https://doi.org/10.1177/002224297103500302

Kubacki, K., Rundle-Thiele, S., Pang, B., & Buyucek, N. (2015). Minimizing alcohol harm: A systematic social marketing review. *Journal of Business Research, 68*(10), 2214–2222. https://doi.org/10.1016/j.jbusres.2015.03.023

Lee, N. R., & Kotler, P. (2015). *Social marketing: Influencing behaviors for good* (5th ed.). Sage.

Master, Z., & Resnik, D. (2013). Hype and public trust in science. *Science and Engineering Ethics, 19*(2), 321–335. https://doi.org/10.1007/s11948-011-9327-6

McKenzie-Mohr, D. (2011). *Fostering sustainable behavior: An introduction to community-based social marketing* (3rd ed.). New Society Publishers.

McKenzie-Mohr, D. (2022). Selecting behaviors: A critical first step in community-based social marketing. In *The Palgrave encyclopedia of social marketing* (pp. 1–6). Palgrave Macmillan. https://doi.org/10.1007/978-3-030-14449-4_30-1

McKenzie-Mohr, D., Lee, N. R., Schultz, W. P., & Kotler, P. (2012). *Social marketing to protect the environment: What works.* Sage Publications.

Miller, S. (2001). Public understanding of science at the crossroads. *Public Understanding of Science, 10*(1), 115–120. https://journals.sagepub.com/doi/abs/10.3109/a036859

Obermiller, C. (1995). The baby is sick/the baby is well: A test of environmental communication appeals. *Journal of Advertising, 24*(2), 55–70. https://doi.org/10.1080/00913367.1995.10673476

Pezzullo, P., & Cox, R. (2021). *Environmental communication and the public sphere* (6th ed.). Sage.

Rodriguez-Sanchez, C., & Sancho-Esper, F. M. (2016). Alcohol regulation, communication strategies and underage alcohol consumption in Spain: Implications for social marketing. *Journal of Social Marketing, 6*(4), 390–411. https://doi.org/10.1108/JSOCM-08-2015-0056

Scheufele, D. (2013). Communicating science in social settings. *Proceedings of the National Academy of Science of the United States of America (PNAS), 110*(3), 14040–14047. https://doi.org/10.1073/pnas.1213275110

Stoutenborough, J., & Vedlitz, A. (2014). The effect of perceived and assessed knowledge of climate change on public policy concerns: An empirical comparison. *Environmental Science & Policy, 37*, 23–33. https://doi.org/10.1016/j.envsci.2013.08.002

Sturgis, P., & Allum, N. (2004). Science in society: Re-evaluating the deficit model of public attitudes. *Public Understanding of Science, 13*(1), 55–74. https://doi.org/10.1177/0963662504042690

Thaler, R. H., & Sunstein, C. R. (2009). *Nudge: Improving decisions about health, wealth, and happiness.* Penguin Books.

Truong, V. D., & Dang, N. V. H. (2016). Reviewing research evidence for social marketing: Systematic literature reviews. In K. Kubacki & S. Rundle-Thiele (Eds.), *Formative research in social marketing: Innovative methods to gain consumer insights* (pp. 183–250). Springer Nature Singapore. https://doi.org/10.1007/978-981-10-1829-9_11

World Economic Forum. (2011). *Water security: The water–energy–food–climate nexus.* Island Press. http://www3.weforum.org/docs/WEF_WI_WaterSecurity_WaterFoodEnergyClimateNexus_2011.pdf

Yao, Q. (2008). Exploring the social dynamics in the U.S. democracy: Presidential and public opinions about, and media coverage of, environmental issues. Doctoral dissertation, University of South Carolina.

Yao, Q., Chang, C.-F., Joshi, P., & McDonald, C. (2024). Climate change versus the water-energy-food nexus: The oldness or newness of the scientific issue as a factor in the deficit model and the hierarchy of response model. *Environment, Development and Sustainability, 26*, 823–840. https://doi.org/10.1007/s10668-022-02735-3

Yao, Q., Joshi, P., Chang, C.-F., McDonalds, C., Tran, J., Wheeler, W., & Hou, S. (2018). Advocating a new approach to governing water, energy, and food security: Testing the effects of message inoculation and conclusion explicitness in the case of the WEF nexus. *Journal of International Crisis and Risk Communication Research, 1*(1), 111–133. https://jicrcr.com/index.php/jicrcr/article/view/6/6

Yao, Q., Liu, Z., & Stephens, L. F. (2020). Exploring the dynamics in the environmental discourse: The longitudinal interaction among public opinion, presidential opinion, media coverage, policymaking in 3 decades and an integrated model of media effects. *Environmental Systems and Decisions, 40*(1), 14–28. https://doi.org/10.1007/s10669-019-09746-y

CHAPTER 4

The Audience Data on the WEF Nexus: The National College Student Science Literacy Survey

College students represent the future and hope. Their understanding of, and attitude toward, science and environmental issues, including the concept of the water-energy-food (WEF) nexus, determines whether solutions to those issues can be forwarded in the next decades. To assess college students' opinions on those issues, regular monitoring through tools such as representative-sample surveys is necessary. Since no national survey of American college students on science and environmental issues has been identified yet, the National College Student Science Literacy Survey (NCSSLS), conducted in 2016–2018, provides valuable data to understand college students' science literacy and environmental knowledge and attitudes as well as their responses to a persuasive stimulus of message levels and factors, especially regarding the concept of the WEF nexus.

The survey was piloted on a university campus and then conducted in two rounds across the U.S. The first round focused on the representativeness of the sample and aimed to present an accurate cross-sectional description of those college students' understanding of, and support for, science and environmental issues. The second round, in which participants were randomly assigned to 11 WEF-nexus message groups and the no-message control group, was essentially a survey experiment designed to test the hypotheses and the research question on message levels and factors proposed in the following chapters.

Q. J. Yao, *Marketing the Water-Energy-Food (WEF) Nexus*,
https://doi.org/10.1007/978-3-031-93148-2_4

1 The Pilot Round

The pilot round was conducted from December 1 to 24, 2016, through SurveyMonkey.com to students at a mid-sized public university in Texas, collecting 524 valid responses. Of these valid responses (between 18 and 72 years old), 19% were from freshmen, 16% sophomores, 20% juniors, 33% seniors, 5.4% master's students, 2.7% doctoral students, and 2.7% all-but-dissertation (ABD) students. The survey took respondents an average of 15 minutes to complete.

This round was embedded with only four versions of the WEF-nexus message as the survey-experimental stimulus: the open-ended one-sided version, the open-ended two-sided version, the closed-ended one-sided version, and the closed-ended two-sided version. The four versions, along with a blank page for the control group, were randomly shown to the respondents after they answered the initial questions assessing their attitudes toward science, science literacy, science professionals, and other issues. The purpose of this pilot study was to test the questionnaires for the formal rounds of the survey and the versions as the single-message stimulus in the second round. The results of the pilot study were published by Yao et al. (2018).

2 The First Round

The first round was conducted from April 3 to May 1, 2017, via SurveyMonkey.com, reaching 4103 students randomly selected from 1501 universities and colleges across the U.S. The list of universities and colleges was taken from the ten 2016 rankings of U.S. colleges by the U.S. News & World Report (national and regional—north, south, west, and midwest—universities and liberal arts colleges. The ranking categories were changed later).

2.1 *Sampling and Administration*

To take a sample representative of the U.S. college student population, six types of information about the 1501 universities and colleges from these ten ranking lists were collected: rank, school name, total enrollment, undergraduate enrollment, and undergraduate male and female percentages. To calculate the quota of students for each university or college, a percentage was calculated as its enrollment in 2017 over the total enrollment for those 1501 universities and colleges in that same year, weighting its graduate/undergraduate and undergraduate male/female ratios. For instance, Princeton, based on the quota calculation, was allocated 2.4 graduate students, 2.5 male undergraduate students, and 2.3 female undergraduate students, 7.1 students in total (rounded down to seven students) for the sample. Williams College's quota was calculated as 0.05 graduate students, 0.9 male undergraduate students, and 0.9 female undergraduate students, equal to 1.8 students in total (rounded up to two).

Research assistants then accessed the online directories of the universities or colleges (some colleges combined their online directories with the search boxes on their homepages) and randomly searched for students to meet those quotas. To randomly select a student, research assistants used an online random name generator (https://www.behindthename.com/random/) and generated a name (encompassing all genders and cultures) for each search (a last name was also generated with the first name). The first student name that appeared upon the search (the same as, or similar to, the randomly generated name for the search) was taken into the sample for that university or college. If no name appeared, the first name and the last name were searched separately. If still no student name popped up, a new name was then generated and searched again in the same way.

For universities or colleges without online directories, requests were sent to their registrars to draw a student sample to meet their quota. If the registrars did not respond, students' names and emails were obtained from webpages of their student clubs or publications.

For all invited participants, 11 iPad Air 2 tablets were provided as incentives for participating in this round (please see the Appendix for the invitation email). A total of 676 valid responses were collected from 235 U.S. universities and colleges. Without including message stimuli, this was mainly a cross-sectional measurement of college students' science literacy and opinions on science and environmental issues, including the WEF nexus. This is arguably the most representative sample of U.S. college students so far and also provides valuable information on their media use and lifestyle.

2.2 *Variables*

Keeping up with Science News. The question asked respondents "How much do you keep up with news about science?" with a ten-point scale ranging from "not at all" to "a lot" ($N = 670$, $Min = 1$, $Max = 10$, $M = 5.41$, $SD = 2.60$).

Science Literacy. The survey used the nine-item science literacy test developed by the National Science Board in its annual S&E Indicator survey for decades (National Science Board, 2016) to measure college students' science literacy. Six of the true/false questions were true and three of them were false, and respondents earned one point if answering one question correctly. So the scores for the science literacy test ranged from 0 to 9 and the U.S. college students on average scored a B (80.7%) on the test ($N = 676$, $Min = 0$, $Max = 9$, $M = 7.27$, $SD = 1.62$).

Attitudes Toward Science and Technology. The survey also asked students if science and technology overall had made most people's lives easier or more difficult. The scale ranged from 0 = extremely easier to 100 = extremely more difficult. College students were generally prone to say that science and technology had helped human life ($N = 665$, $M = 30.69$, $SD = 1.62$).

Subjective WEF-Nexus Knowledge 1. The question asked respondents how much they knew about the WEF nexus by selecting a given answer ranging from 1 (= I have never heard about it) to 10 (= I am very knowledgeable about it). The question was asked at the beginning of the survey (being the fourth question). The WEF nexus was confirmed to be a fairly new concept for college students ($N = 670$, $M = 1.77$, $SD = 1.62$).

Trust in Scientists and Others. The battery of seven questions asked respondents how much they trusted (1) scientists, (2) journalists, (3) professors, (4) family, (5) neighbors, (6) friends, and (7) a typical person in society. All questions were measured with a scale ranging from 0 (= not at all) to 7 (= extremely). The seventh question was an established measurement of "social trust" (Yao et al., 2019). College students generally trusted family ($N = 667$, $M = 5.85$, $SD = 1.37$), friends ($N = 666$, $M = 5.20$, $SD = 1.30$), scientists ($N = 667$, $M = 5.00$, $SD = 1.36$), and professors ($N = 666$, $M = 4.90$, $SD = 1.25$) more than they trusted neighbors ($N = 666$, $M = 3.50$, $SD = 1.31$), journalists ($N = 667$, $M = 3.11$, $SD = 1.27$), and a typical person in society ($N = 669$, $M = 2.90$, $SD = 1.09$).

Perceived Believability of Science News. The question asked respondents to rate how believable the science news from the mass media was on a ten-point scale ranging from "not believable at all" to "completely believable." College students' perception of the believability of science news was in the middle ($N = 646$, $Min = 1$, $Max = 10$, $M = 5.33$, $SD = 1.91$).

Objective WEF-Nexus Knowledge. To measure college students' actual knowledge of the WEF nexus, the first round included eight multiple-choice questions (all with four choices) about the facts related to the WEF nexus, including which section of production consumed the largest share of water globally and in the U.S. respectively, which section of production consumed 30% of the global energy, whether the WEF-nexus concept had been widely adopted, what the impact of bioenergy crops was, what the WEF-nexus concept promoted, what the relationship between the WEF nexus and climate change was, and what percentage of food was thrown uneaten in the U.S. Answering a question correctly earned 1 point; so, the scale ran from 0 to 8 ($N = 676$, $M = 2.58$, $SD = 1.68$).

Subjective WEF-Nexus Knowledge. At the end of the survey, after measuring the objective WEF-nexus knowledge, respondents were asked to rate on a Likert scale (ranging from "strongly disagree" = 1 to "strongly agree" = 5 with neutral = 3) about the statement "I consider myself well informed about issues in the water-energy-food nexus" ($N = 638$, $M = 2.22$, $SD = 1.03$). While multiple-item measures were used whenever possible as suggested by traditional wisdom, single-item measurements were also used when they were more convenient, given that research showed no difference in effect sizes between multi- and single-item measurements, at least in advertising research (Ang & Eisend, 2018).

WEF-Nexus Attitude. A battery of six questions was used to measure college students' supportive attitude toward the WEF-nexus concept, which asked the respondents how much they agreed that it was (1) important; (2) urgent; (3) useful to treat the water, energy, and food security as a nexus; and whether (4) the problems in the WEF nexus were serious; (5) the WEF-nexus concept was just another new bottle for old wines in science (reversely coded); and (6) schools and colleges should teach WEF nexus so that the next generation could understand the relationships among the security of those three resources. The questions were measured with a five-point scale ranging from "strongly disagree" (= 1) to "strongly agree" (= 5). The scale value was the sum of the six items (ranging from 6 to 30), and the scale reliability was decent ($N = 629$, Cronbach's $\alpha = 0.84$, $Min = 6$, $Max = 30$, $M = 21.6$, $SD = 3.67$).

WEF-Nexus Behavioral Intention. A battery of five questions measured college students' behavioral intentions to support the WEF-nexus concept, which included how much they agreed that they would (1) help promote the WEF nexus, (2) help solve problems involved in the WEF nexus, (3) donate to support promoting the WEF nexus, (4) tell people about the WEF nexus, and (5) urge people to support the WEF nexus. Those questions were also measured with the Likert scales ranging from "strongly disagree" (= 1) to "strongly agree" (= 5), with the sum of the five items as the scale value, and the scale reliability was ideal (N = 632, Cronbach's α = 0.88, *Min* = 5, *Max* = 25, M = 15.12, SD = 3.54). The indices of the WEF-nexus attitudes and behavioral intentions were adopted from Gastil and Xenos (2010) and revised.

Environmental Behavioral Intention. Nine questions were used to measure how likely the respondents would take actions to support the environment, including (1) reducing food waste; (2) eating less meat; (3) using less water; (4) buying less stuff and buying reuse, recycle products; (5) stopping drinking bottled water; (6) reducing energy consumption; (7) buying Energy-Star products; (8) traveling with mass-transit or bicycle; and (9) buying an electronic or fuel-efficient car when needed a car. The questions were measured with an eight-point scale ranging from 0 (= not at all) to 7 (= extremely likely). The sum of the nine items was taken as the value of the scale, which was reliable (N = 614, Cronbach's α = 0.88, *Min* = 0, *Max* = 56, M = 47.2, SD = 13.06).

Religiosity. Two questions measured students' religiosity. The first, on a scale from 1 = never to 6 = more than once a week, asked the students how often they attended religious services aside from weddings and funerals. The second asked the students to rank on a scale (0 = not important at all to 100 = extremely important) about the importance of God(s) in their lives (both items were standardized to reduce the difference between the units). The religiosity scale (the sum of the two standardized items) performed well (N = 578, Cronbach's α on standardized items = 0.78, *Min* = −2.61, *Max* = 3.01, M = 0.08, SD = 1.80). Students' religions were also recorded (Protestant, 10.1%; Roman Catholic, 17%; Mormon, 1.3%; Orthodox, 0.5%; Jewish, 3%; Muslim, 2.4%; Buddhist, 1.7%; Hindu, 2.5; Atheist, 11.3%; Agnostic, 8.2%; Nothing in Particular, 15.5%; Christian, 21.6; Unitarian, 0.6%; Jehovah's Witness, 0.2%; Other, 4.1%; for other, specified were belief with no specific religion, Christian Buddhist, Hindu Buddhist, LaVeyan Satanism, People, Seventh-Day Adventist, Shintoism & Christianity, Sikh, Traditional Navajo and Christianity, and prefer not to say).

Happiness. The survey asked respondents "Taking things together, how happy would you say you are?" measured on a ten-point scale ranging from "not happy at all" to "completely happy." College students were overall fairly happy with their lives (N = 672, *Min* = 1, *Max* = 10, M = 7.40, SD = 1.81).

Health-Status Satisfaction. Respondents were asked to rate their state of health on a scale from 0 (= extremely poor) to 100 (= extremely good). College students generally felt healthy ($N = 670$, $M = 78.35$, $SD = 16.90$).

Life Satisfaction. Respondents were asked to rate how satisfied they were with their life as a whole by choosing from a ten-point scale from 1 (= completely dissatisfied) to 10 (= completely satisfied). College students were roughly satisfied with their lives as a whole ($N = 669$, $M = 7.36$, $SD = 1.76$).

Media and Social Media Uses. Two batteries of questions measured college students' media use and social media use. For the media use questions, respondents were asked to rate how often they learn news or information about current events in the past six months through (1) newspapers (print); (2) television; (3) radio; (4) magazines (print); (5) books (print); (6) news websites (including websites of newspapers, televisions, radios, etc.); (7) news apps (including apps of newspapers, televisions, radios, etc.); (8) talking to family, friends, or other people face to face; (9) talking to family, friends, or other people on the phone or mobile calling apps; (10) emails (including email newsletters); and (11) social media (Facebook, Twitter, etc., including social media accounts of mass media) on a nine-point scale ranging from "not at all" (= 9) to "multiple times a day" (= 1; all items reversely coded). College students used such communication channels as, starting from the most often-used, (1) social media ($N = 662$, $M = 7.66$, $SD = 2.02$); (2) talking face to face ($N = 658$, $M = 7.15$, $SD = 1.81$); (3) news websites ($N = 659$, $M = 6.58$, $SD = 2.28$); (4) talking on phone or mobile ($N = 658$, $M = 6.20$, $SD = 2.29$); (5) news apps ($N = 658$, $M = 5.80$, $SD = 2.88$); (6) emails ($N = 658$, $M = 5.56$, $SD = 2.80$); (7) Television ($N = 655$, $M = 5.00$, $SD = 2.57$); (8) radio ($N = 656$, $M = 4.68$, $SD = 2.76$); (9) books (print; $N = 657$, $M = 4.03$, $SD = 2.41$); (10) newspapers (print; $N = 651$, $M = 3.22$, $SD = 2.13$); (11) magazines (print; $N = 657$, $M = 3.14$, $SD = 2.05$).

For the social media questions, respondents were asked how often they posted, commented, or shared news or current events through (1) Facebook, (2) Twitter, (3) LinkedIn, (4) Instagram, (5) Snapchat, (6) personal blog, (7) personal website, and (8) other social media (specify in a given text box) on an eight-point scale ranging from "I don't have an account" (= 1), "about once several months" (= 2), to "multiple times a day" (= 8). College students used such social media as, in the order of favorability, (1) Facebook ($N = 661$, $M = 3.35$, $SD = 2.07$), (2) Snapchat ($N = 658$, $M = 2.91$, $SD = 2.21$), (3) Instagram ($N = 661$, $M = 2.47$, $SD = 1.78$), (4) Twitter

($N = 657$, $M = 2.32$, $SD = 1.92$), (5) LinkedIn ($N = 661$, $M = 1.78$, $SD = 1.08$), (6) other social media ($N = 633$, $M = 1.63$, $SD = 1.52$), (7) personal blog ($N = 660$, $M = 1.40$, $SD = 1.01$) and (8) personal website ($N = 663$, $M = 1.39$, $SD = 1.10$). Specified other social media included: Reddit (listed 18 times), Tumblr (7 times), Pinterest (6 times), WhatsApp (3 times), WeChat (3 times), Imgur (2 times), Slack (2 times), Google+ (1 time), NeoGaf (1 time), 4chan (1 time), Yammer (1 time), YouTube (2 times), Soundcloud (1 time), Tapatalk (1 time), Telegram (1 time), Weibo (1 time), Spotify (1 time), Pandora (1 time), Apple Music (1 time), neighbor apps (1 time), and traditional internet forums (1 time).

Other demographic and socioeconomic variables such as age (also serving as a question to screen out participants younger than 18 years old. $N = 676$, $Min = 18$, $Max = 63$, $M = 23.97$, $SD = 7.33$), gender ($N = 638$, male = 44.8%, female = 54.1%, other = 1.1%; for the "other" category, "agender," "genderqueer," "non-binary," and "n/a" were specified), school classification (from freshmen to ABD; $N = 640$, freshmen = 14.1%, sophomores = 15.6%, juniors = 18.6%, seniors = 24.7%, master students = 13.8%, doctoral students = 7.7%, doctoral candidate/ABDs = 2.2%, and other = 3.4%; for "other," recent graduates, second-degree pursuers, and post-graduate non-degree pursuers were specified. Recoded into a scale variable with "other" coded into "seniors" or "master students" based on their specifications to measure education), GPA (self-reported; $N = 612$, $Min = 1.80$, $Max = 4.80$, $M = 3.45$, $SD = 0.46$), areas of studying (arts and humanities, 13.3%; social sciences, 35.9%; natural sciences, 19.4%; medical sciences, 14.1%; engineering, 17.1%; other/undecided, 0.2%), program enrolled ($N = 638$; on-campus program with all courses delivered in person, 63.6%; online program with most courses delivered in person but some online, 27.4%; online program with all courses delivered through internet, 4.1%; online program with most courses delivered through internet but some required in person, 2.7%; other, 2.2%; for other, hybrid or clinic were specified), ethnicities ($N = 667$, White, 64.8%; Black, 8.1%; Asian 14.5%; Hispanic, 12.6%; Pacific Islander, 0.7%; Native American, 1.9%; Other 1.3%), annual family income ($N = 635$; less than \$10,000, 8.7%; \$10,000–20,000, 6.5%; \$20,000–30,000, 6.6%; \$30,000–40,000, 5.7%; \$40,000–50,000, 6.0%; \$50,000–75,000, 11.0%; \$75,000–100,000, 9.4%; \$100,000–150,000, 13.5%; \$150,000 or more, 10.4%; don't know, 22.2%—treated as missing value when analyzing income), partisanship ($N = 634$; Republican, 11.2%; lean toward Republican, 9.3%; Independent, 13.1%; lean toward Democratic, 18.9%; Democratic, 27.4%; no preference, 18.1%; other parties, 1.9%; for "other parties," Green, Conservative libertarian, Democratic socialist, Libertarian, and "not an American" were specified), and ideology ($N = 635$; very conservative, 3.8%; conservative, 13.2%, moderate, 27.2%; liberal, 27.6%; very liberal, 18.9%; don't know/refused, 9.3%, coded as missing when ideology analyzed as a scale) were also measured and controlled to ensure the associations detected were substantial.

2.3 *Gauging College Students' Science Literacy, WEF-Nexus Opinions, and Predispositions*

Data collected from the first round, based on the most representative sample available by far, provided a cross-sectional description of American college students' science literacy, factors contributing to it, relationships among objective and subjective knowledge, attitudes, and behavioral intentions toward the WEF nexus and environmental protection, and the relationship between religious beliefs and attitudes toward science and technology.

Factors Contributing to College Students' Science Literacy
What factors contribute to college students' science literacy, when potentially confounding socio-demographic factors were controlled? To answer this question, science literacy was regressed on all related predictors to explore those meaningful contributors, and the regression model was finalized after excluding the clearly insignificant independent variables (see Table 4.1).

The model confirmed some of our educated guesses on the contributors to college students' science literacy. For instance, students who studied natural science, medical science, or engineering scored higher on the science literacy test ($B = 0.46$, $SE = 0.15$, $\beta = 0.15$, $p = 0.002$), compared to those who studied social sciences and arts and humanities. Family income was positively correlated to a student's science literacy ($B = 0.10$, $SE = 0.03$, $\beta = 0.17$, $p < 0.001$) and perhaps overall academic performance. Students who kept up with science news also scored higher on the science literacy test ($B = 0.07$, $SE = 0.03$, $\beta = 0.12$, $p = 0.03$).

The model also revealed positive or negative predictors of college student science literacy that were not initially apparent and therefore directed supporters and advocates to ways of improving college science literacy. For instance, among ethnic groups, Black students showed a need for special support to increase science literacy ($B = -0.89$, $SE = 0.30$, $\beta = -0.16$, $p = 0.003$). Also, students of Muslim ($B = -2.45$, $SE = 0.70$, $\beta = -0.16$,

Table 4.1 Predictors of college students' science literacy

IVs		*B*	*SE*	*ß*	*p*
Block 1	Constant	4.55	0.78		<0.001[c]
	Age	0.01	0.01	0.73	0.21
	Education	0.08	0.04	0.10	0.065
	Gender (*M* = 1)	0.28	0.16	0.09	0.08
	Area of Study (science, medicine, or engineering = 1, else = 0)	0.46	0.15	0.15	0.002[b]
	Family income	0.10	0.03	0.17	<0.001[c]
	Liberal ideology	0.17	0.07	0.12	0.02[a]
	White (= 1, else = 0)	0.21	0.26	0.06	0.42
	Black (= 1, else = 0)	−0.89	0.30	−0.16	0.003[b]
	Hispanic (= 1, else = 0)	0.47	0.26	0.11	0.07
	Asian (= 1, else = 0)	−0.15	0.29	−0.03	0.61
	Protestant (= 1, else = 0)	−0.36	0.25	−0.07	0.15
	Roman Catholic (= 1, else = 0)	−0.15	0.21	−0.04	0.45
	Muslim (= 1, else = 0)	−2.45	0.70	−0.16	<0.001[c]
	General Christian (= 1, else = 0)	−0.52	0.20	−0.14	0.008[b]
	Atheist (= 1, else = 0)	−0.19	0.23	−0.04	0.42
Block 2	GPA	0.31	0.17	0.09	0.06
	Keeping up with science news	0.07	0.03	0.12	0.03[a]
	Trust in journalists	−0.06	0.06	−0.05	0.28
	Attitude toward S&T	−0.004	0.002	−0.08	0.075
	Subjective health status	−0.006	0.004	−0.07	0.15
	Enrolling in online program (= 1, else = 0)	−0.17	0.27	−0.03	0.53
Block 3	Newspaper info use	0.06	0.04	0.08	0.10
	TV info use	−0.06	0.03	−0.09	0.054
	Book (print) info use	−0.08	0.03	−0.12	0.016[a]
	News app info use	0.08	0.03	0.14	0.002[b]
Model info	$R^2 = 0.32$; *Adj.* $R^2 = 0.27$				

[a]Stands for significance at the 0.05 level
[b]Stands for significance at the 0.01 level
[c]Stands for significance at the 0.001 level

$p < 0.001$) and general Christian (not self-identifying as Roman Catholics or Protestants; $B = -0.52$, $SE = 0.20$, $ß = -0.14$, $p = 0.008$) demonstrated lower levels of scientific knowledge. For college students, while reading books (print) for information or current events decreased science literacy ($B = -0.08$, $SE = 0.03$, $ß = -.12$, $p = 0.016$), using news apps for such purpose increased their science literacy ($B = 0.08$, $SE = 0.03$, $\beta = 0.14$, $p = 0.002$). Compared with college students who leaned toward being conservative, those who leaned toward being liberal demonstrated higher science literacy ($B = 0.17$, $SE = 0.07$, $\beta = 0.12$, $p = 0.02$).

The stereotypes that male ($B = 0.28$, $SE = 0.16$, $\beta = 0.09$, $p = 0.08$) or Asian ($B = -0.15$, $SE = 0.29$, $\beta = -0.03$, $p = 0.61$) college students held an advantage in learning science (Lee, 1998; Riegle-Crumb & King, 2010) and that online students ($B = -0.17$, $SE = 0.27$, $\beta = -0.03$, $p = 0.53$) might be disadvantageous in learning were found unwarranted.

Subjective vs. Objective WEF-Nexus Knowledge and Their Impacts on WEF-Nexus Attitudes, Behavioral Intentions

Knowledge as a concept is especially important in science or environmental communication because science or environmental issues usually require special knowledge or even expertise to assess (Fischer, 2000). Even when coping with COVID-19 (Hesaraki et al., 2020) or other healthcare issues (Kanu et al., 2021), communication campaigns usually start from the knowledge and gradually achieve attitudinal or behavioral goals.

Knowledge can be measured subjectively with respondents' self-report of their knowledge of the subject matter or objectively through respondents' taking tests about the subject matter. The former is more popular. The correlation between subjective measurements and objective measurements, however, has been found weak, or even nonexistent in previous studies, meaning people's perceptions of their knowledge and their real knowledge of a topic can be distant. Also, subjective knowledge is found more associated with related behaviors compared to objective knowledge (Han, 2019; Lee et al., 2022). The disparity between subjective knowledge and objective knowledge could result from people's general pattern of overestimating self and superficial learning on an issue that is prevalent in the context around them, which falsely makes people feel informed (Lee et al., 2022). On a new topic like the WEF nexus, the contribution of superficial learning is considered lower, because many people haven't heard about this term.

The first round data contain variables to examine the relationships among subjective and objective knowledge, attitudes, and behavioral intentions. Those variables are, specifically, science literacy (objective knowledge of science), subjective WEF-nexus knowledge 1, subjective WEF-nexus knowledge, objective WEF-nexus knowledge, WEF-nexus attitudes, WEF-nexus behavioral intentions, and environmental behavioral intentions. Their zero-order and partial (controlling age, gender, education, family income, and ideology) correlations were analyzed (see Table 4.2).

Science literacy was uncorrelated (in both the zero-order and partial tests) to the two subjective measurements of WEF-nexus knowledge and the behavioral intention toward the WEF nexus. It was, however, correlated to objective WEF-nexus knowledge (zero-order: $r = 0.27$, $N = 676$, $p < 0.001$; partial: $r = 0.27$, $df = 411$, $p < 0.001$) and WEF-nexus attitudes (zero-order: $r = 0.19$, $N = 629$, $p < 0.001$; partial: $r = 0.18$, $df = 411$, $p < 0.001$). Its zero-order correlation with environmental behavioral intention (zero-order: $r = 0.11$, $N = 614$, $p = 0.009$; partial: $r = 0.09$, $df = 411$, $p = 0.08$) disappeared when demographic variables were controlled. Both subjective measurements of WEF-nexus knowledge were correlated to each other, although not strongly (zero-order: $r = 0.31$, $N = 632$, $p < 0.001$; partial: $r = 0.28$, $df = 411$, $p < 0.001$), and to the objective measurement of WEF-nexus knowledge (for subjective WEF-nexus knowledge 1, zero-order: $r = 0.22$, $N = 670$, $p < 0.001$; partial: $r = 0.22$, $df = 411$, $p < 0.001$. For subjective WEF-nexus knowledge, zero-order: $r = 0.18$, $N = 638$, $p < 0.001$; partial: $r = 0.21$, $df = 411$, $p < 0.001$), WEF-nexus attitudes (for subjective WEF-nexus knowledge 1, zero-order: $r = 0.21$, $N = 623$, $p < 0.001$; partial: $r = 0.20$, $df = 411$, $p < 0.001$. For subjective WEF-nexus knowledge, zero-order: $r = 0.19$, $N = 628$, $p < 0.001$; partial: $r = 0.17$, $df = 411$, $p < 0.001$), WEF-nexus behavioral intention (for subjective WEF-nexus knowledge 1, zero-order: $r = 0.28$, $N = 626$, $p < 0.001$; partial: $r = 0.24$, $df = 411$, $p < 0.001$. For subjective WEF-nexus knowledge, zero-order: $r = 0.38$, $N = 631$, $p < 0.001$; partial: $r = 0.35$, $df = 411$, $p < 0.001$), as well as environmental behavioral intention (for subjective WEF-nexus knowledge 1, zero-order: $r = 0.19$, $N = 608$, $p < 0.001$; partial: $r = 0.13$, $df = 411$, $p = 0.009$. for subjective WEF-nexus knowledge, zero-order: $r = 0.18$, $N = 611$, $p < 0.001$; partial: $r = 0.19$, $df = 411$, $p < 0.001$).

Objective WEF knowledge also clearly correlated to WEF-nexus attitudes (zero-order: $r = 0.35$, $N = 629$, $p < 0.001$; partial: $r = 0.31$, $df = 411$, $p < 0.001$), WEF-nexus behavioral intention (zero-order: $r = 0.20$, $N = 632$, $p < 0.001$; partial: $r = 0.17$, $df = 411$, $p < 0.001$), and environmental behavioral intention (zero-order: $r = 0.22$, $N = 614$, $p < 0.001$; partial: $r = 0.19$, $df = 411$, $p < 0.001$). Again, perhaps because WEF nexus was a new topic, and respondents' subjective perception of their knowledge on the topic had not been so distant from their real knowledge (Lee

Table 4.2 Zero-order and partial correlations between variables of WEF-Nexus knowledge, attitude, and behavioral intentioans

		Subjective knowledge 1	*Subjective knowledge*	*Objective knowledge*	*Attitude*	*Behavioral intention*	*Environmental behavioral intention*
Science literacy	Zero-order	0.05, 0.17, 670	−0.04, 33, 638	0.27, <0.001, 676[c]	0.19, <0.001, 629[c]	0.05, 0.22, 632	0.11, 0.009, 614[b]
	Partial	0.09, 0.08, 411	0.002, 0.96, 411	0.27, <0.001, 411[c]	0.18, <0.001, 411[c]	0.06, 0.22, 411	0.09, 0.08, 411
Subjective knowledge 1	Zero-order		0.31, <0.001, 632[c]	0.22, <0.001, 670[c]	0.21, <0.001, 623[c]	0.28, <0.001, 626[c]	0.19, <0.001, 608[c]
	Partial		0.28, <0.001, 411[c]	0.22, <0.001, 411[c]	0.20, <0.001, 411[c]	0.24, <0.001, 411[c]	0.13, 0.009, 411[b]
Subjective knowledge	Zero-order			0.18, <0.001, 638[c]	0.19, <0.001, 628[c]	0.38, <0.001, 631[c]	0.18, <0.001, 611[c]
	Partial			0.21, <0.001, 411[c]	0.17, <0.001, 411[c]	0.35, <0.001, 411[c]	0.19, <0.001, 411[c]
Objective knowledge	Zero-order				0.35, <0.001, 629[c]	0.20, <0.001, 632[c]	0.22, <0.001, 614[c]
	Partial				0.31, <0.001, 411[c]	0.17, <0.001, 411[c]	0.19, <0.001, 411[c]
Attitude	Zero-order					0.54, <0.001, 623[c]	0.44, <0.001, 605[c]
	Partial					0.54, <0.001, 411[c]	0.41, <0.001, 411[c]

(*continued*)

Table 4.2 (continued)

		Subjective knowledge 1	*Subjective knowledge*	*Objective knowledge*	*Attitude*	*Behavioral intention*	*Environmental behavioral intention*
Behavioral intention	Zero-order						0.53, <0.001, 605[c]
	Partial						0.51, <0.001, 411[c]

Subjective knowledge 1 was measured with: "How much do you know about water-energy-food nexus?" Subjective knowledge was measured with respondents' degree of agreement on: "I consider myself well informed about issues in the water-energy-food nexus"

Entries are Pearson's *r*, *p*-value, and degree of freedom

For partial correlations, age, gender ($M = 1$, else = 0), education, family income, and ideology were controlled. Data are from the NCSSLS first round

[a]Stands for significance at 0.05 level
[b]Stands for significance at 0.01 level
[c]Stands for significance at 0.001 level

et al., 2022). According to Lee and Preacher's (2013) calculator to test the difference between correlations with one variable in common, objective WEF knowledge showed stronger correlations with WEF-nexus attitudes than both subjective WEF-nexus knowledge 1 (for the partial coefficients, $p = 0.06$) and subjective WEF-nexus knowledge (for the partial coefficients, $p = 0.018$). Subjective WEF-nexus knowledge, however, showed a stronger correlation to WEF-nexus behavioral intention than objective WEF-nexus (for the partial coefficient, $p = 0.002$) and even subjective WEF-nexus knowledge 1 (for the partial coefficient, $p = 0.048$), signaling that it was a better measurement of subjective knowledge than subjective WEF-nexus knowledge 1.

As expected, WEF-nexus attitude is moderately correlated to WEF-nexus behavioral intention (zero-order: $r = 0.54$, $N = 623$, $p < 0.001$; partial: $r = 0.54$, $df = 411$, $p < 0.001$) and further the environmental behavioral intention (zero-order: $r = 0.44$, $N = 605$, $p < 0.001$; partial: $r = 0.41$, $df = 411$, $p < 0.001$), with the latter two also moderately correlated (zero-order: $r = 0.53$, $N = 605$, $p < 0.001$; partial: $r = 0.51$, $df = 411$, $p < 0.001$). As a new topic with minimal prior knowledge, the relationships between the knowledge, attitude, and behavioral intention regarding the WEF nexus were clearly identifiable (Yao et al., 2024).

The Dialogue Between the Gods and Science Among College Students

The relationship between religion and science has attracted attention for centuries. Religious followers have defended religion from being ridiculed as backward and anti-science. They have provided evidence that religiosity has been the major driver of scientific advancement, arguing that most founders of modern science frameworks and Nobel laureates in science are religious followers and that modern scientific knowledge was brought to developing areas of the world, such as Africa, Eastern Asia, and South America, largely by the missionaries during the sixteenth–nineteenth centuries. One of their favorite stories is Isaac Newton's devotion in his later life to proving the existence of God.

Many scholars also maintain that religion is complementary to science. They hold that (1) religion boosters science, (2) science needs religion to solve some ultimate questions, and (3) science enhances religion. George Salib, professor of Arabic and Islamic science at Columbia University, for instance, argues that Islamic culture had boosted a golden age of science, through such scientific advancements as developing the experiment research method and diagramming the structural, nervous, and circulatory systems in the human body, during Europe's Dark Ages (Kruglinski, 2007). Alfred Wallace, the proposer of the theory of natural selection as the impetus of evolution, also believes that the development of the human brain could only be explained when a concept of divine force was introduced (Annas, 2006). Nobel Laureate in Physiology or Medicine John Eccles (1989) notes that the self and soul of a human being could only be implanted into a human fetus by a "supernatural creation." Stephen Hawking (1988) also pointed out that even if one day physicians can construct a unified theory of the universe, they will still be confronted with questions about the purposes of human existence and the nature of God.

But other people still feel the tension between religion and science. In his lecture at MIT, Muslim scholar Seyyid Nasr mentioned that the evening when Muslim children learned from school that water was composed of oxygen and hydrogen they stopped to say their prayers, which described water as a divine purifier (Nasr, n.d.). Discussion of the conflict between religion and science, as former president Jimmy Carter (2005) put it, is one of the ancient and persistent debates, especially in the U.S. Carter, a strong Christian since his childhood and a science graduate, was frustrated when he internally experienced the conflict between his beliefs and his scientific research. He later had to find a way to completely separate science and religion: religion should not intervene in science classrooms and neither of them can prove the other (Carter, 2005). Bertrand Russell (1961) did a great job in elucidating the Roman Catholic Orthodox's suppression of science, which he believed to be acute in the sixteenth century and continued down to nowadays. The intrinsic reason is that scientific advancement attacks dogmas widely accepted in the religious world, while religion needs to suppress scientific advancement to maintain its own consistency and authority. The discussions on the center of the universe and the theory of evolution were two major cases that Russell (1961) noted.

The intelligent design lawsuit in the U.S. was also a good example to illustrate the tension between science and religion. The lawsuit, in essence, as Judge John Jones pointed out in the court's decision, is the antagonism between religion and science, in particular the theory of evolution. George Annas (2006) noted that religious followers have pushed two waves of anti-evolution movements: the 1920s movement outlawing teaching evolution and anything that denied the Bible and the 1980s movement requesting teaching the creationism theory. Judge Jones' decision marked the end of the third wave, which requested teaching intelligent design, a new format of creationism. Lately, a new book features a pastor and a science writer promoting the inclusion of evolution into Christian stories, creating a "Genesis 2.0" (Sideris, 2017, p. 146).

The relationship between religion and science inevitably impacts students' learning of science at schools. The first round also provided data to explore how American college students' religiosity relates to their attitudes toward science and technology. Attitudes toward science, with a higher value standing for a belief that science and technology have made life more difficult for most people, were regressed on all possibly related socio-demographic, religious, and informational variables, with the final model established by excluding non-religion variables that were also far from significant (see Table 4.3).

The model showed that, with possible confounders under control, college students' religiosity was not related to their attitude toward science and technology (S&T, $B = 1.48$, $SE = 1.13$, $\beta = 0.09$, $p = 0.189$). None of the religions or denominations generated a negative attitude toward science and technology; instead, Protestantism clearly generated a favorable attitude toward science and technology ($B = -15.56$, $SE = 5.56$, $\beta = -0.17$, $p = 0.005$). Why Protestantism has such a unique impact on college students' attitudes toward science and technology remains unknown and interesting for further exploration.

Also noticeable was Asian college students' agreement with the statement that science and technology had made life more difficult for most people ($B = 13.95$, $SE = 5.86$, $\beta = 0.17$, $p = 0.018$). Again, the underlying cultural or psychological mechanisms were unclear, but the phenomenon is interesting for continuous examination. Trust in scientists, with no surprise, made college students believe more that science and technology have helped people's lives ($B = -3.12$, $SE = 1.39$, $\beta = -0.15$, $p = 0.025$). Using news apps for current events, interestingly and perplexingly, also made college students believe that science and technology had made life more difficult for most people ($B = 1.08$, $SE = 0.53$, $\beta = 0.11$, $p = 0.044$), perhaps because those tech-savvy readers had learned about how others struggle to use news apps and other information or life-assisting technology.

Table 4.3 The regression of attitude toward science and technology on religiosity and other factors

IVs		*B*	*SE*	*ß*	*p*
Block 1	Constant	42.91	13.47		0.002[b]
	Age	0.10	0.23	0.03	0.675
	Education	−0.42	0.91	−0.03	0.65
	Gender (*M* = 1)	3.73	3.08	0.06	0.226
	Family income	−0.16	0.57	−0.01	0.78
	White (= 1, else = 0)	6.71	5.27	0.11	0.204
	Black (= 1, else = 0)	6.95	6.06	0.07	0.252
	Hispanic (= 1, else = 0)	4.17	5.45	0.05	0.445
	Asian (= 1, else = 0)	13.95	5.86	0.17	0.018[a]
Block 2	Democrat (= 1, else = 0)	2.96	3.33	0.05	0.374
	Trust in scientists	−3.12	1.39	−0.15	0.025[a]
	Trust in professors	2.07	1.46	0.09	0.158
	Life happiness	−1.02	0.84	−0.06	0.224
Block 3	Protestant (= 1, else = 0)	−15.56	5.56	−0.17	0.005[b]
	Roman Catholic (= 1, else = 0)	1.33	4.61	0.02	0.774
	Muslim (= 1, else = 0)	9.36	13.25	0.04	0.480
	General Christian (= 1, else = 0)	−0.75	4.65	−0.01	0.873
	Agnostic (= 1, else = 0)	7.69	5.63	0.08	0.173
	Atheist (= 1, else = 0)	−1.64	5.72	−0.2	0.774
	Religiosity	1.48	1.13	0.09	0.189
Block 4	Keep up with science news	−0.74	0.69	−0.07	0.284
	Talking face-to-face for info	−0.95	0.98	−0.06	0.332
	Talking through calls for info	1.40	0.77	0.11	0.069
	Newspaper info use	0.55	0.76	0.04	0.468
	TV info use	−0.74	0.61	−0.07	0.226
	Social media info use	−1.16	0.78	−0.08	0.141
	News app info use	1.08	0.53	0.11	0.044[a]
Model info	$R^2 = 0.11$; *Adj.* $R^2 = 0.05$				

[a]Stands for significance at the 0.05 level
[b]Stands for significance at the 0.01 level
[c]Stands for significance at the 0.001 level

For the DV, the higher the value, the more negative the attitude toward science and technology

3 The Second Round: Data to Test the WEF-Nexus Message

The second round was administered with the same sample through SurveyMonkey.com, from September 13, 2017 to January 8, 2018 (with email unsubscriptions, 3539 students from about 250 universities and colleges remained in the sample), which took subjects on average about

11 minutes to complete. Seven iPad Air 2 tablets were mailed to randomly selected participants, and follow-up messages with famous quotes on science were sent to all students in the sample on a two- to three-day basis to enhance students' interest in the project and remind them to participate. From 185 universities, 402 valid responses were collected. The subjects aged between 18 and 63 (N = 402, M = 24.4, SD = 7.15). Freshmen accounted for 1 (0.3%), sophomores 57 (15.6%), juniors 79 (21.6%), seniors 101 (27.7%), master's students 53 (14.5%), doctoral students 29 (7.9%), ABDs 12 (3.3%), and other 33 (8.2%). Male, 173 (47.5%); female, 190 (52.2%), other 1 (.3%). 52 (14.3%) students studied arts and humanities, 123 (33.8%) social sciences, 78 (21.4) natural sciences, 44 (12.1%) medical sciences, 61 (16.8) engineering, and 6 (1.6) other (not specified). Among them, 269 (67.1%) took the first round of the survey, 60 (15%) did not, and 72 (18%) did not remember. Subjects declared their religions as follows: Protestant (9.6%), Roman Catholic (17.3%), Atheist (9.9%), Agnostic (10.7%), Nothing in Particular (17.9%), Christian (21.7%), Buddhist (1.9%), Jewish (1.6%), and Muslim (1.6%). For ethnicity, White, 236 (58.7%); Black, 40 (10%), Hispanic, 49 (12.2%), Asian, 58 (14.4%), Native American, 6 (1.5%), and other, 6 (1.5%).

To examine the effects of the message levels and factors with the versions of the WEF-nexus message embedded in the questionnaire, the second round, essentially designed as a survey-experiment, did not emphasize sample representativeness. Most of the variables in the first round appeared in the second round too, except that for three groups of variables. For the media use variables (still measured with 1 = multiple times a day to 9 = not at all, all reversely coded) the media choices were reduced to (1) print media (newspapers, magazines, and books, N = 395, M = 4.14, SD = 2.34); (2) broadcasting (TV, radio, and cables, N = 396, M = 4.77, SD = 2.47); (3) websites, mailing lists, emails, news apps, and so on (N = 395, M = 6.20, SD = 2.45); (4) social media (N = 395, M = 6.55, SD = 2.39); and (5) talking to family, friends, or other people (face to face, calling, etc.; N = 397, M = 5.87, SD = 2.28). Choices for the trust variables were reduced to (1) scientist (N = 397, Min = 1, Max = 7, M = 5.37, SD = 1.30), (2) journalists (N = 396, Min = 1, Max = 7, M = 3.38, SD = 1.35), (3) professors (N = 396, Min = 1, Max = 7, M = 3.38, SD = 1.35), and (4) a typical person in society (N = 397, Min = 1, Max = 7, M = 2.89, SD = 1.09). For GPA, the self-report box was changed to closed choices, ranging from below 1.0 to above 4.0 in increments of 0.5 (N = 365, 1.51–2.0, 0.8%, 2.01–2.5, 3%, 2.51–3, 14%, 3.01–3.5, 31%, 3.51–4.0, 49.9%, 4.0+, 1.4%).

The message versions embedded in the questionnaires were manipulated according to levels and factors for the participants to read first and then provide their responses (knowledge, attitudes, and behavioral intentions). The following are the variables in the second round and will be used to test the hypotheses and research question proposed in the following chapters.

3.1 The Message Variable on the WEF Nexus

Within the second round of the survey, a question with 12 message options was embedded. Subjects were randomly exposed to no message (control group) or one of the 11 versions of the WEF-nexus message that was designed to test the persuasive effect of four levels of media stimuli (agenda, knowledge, attitude, and behavior), message sidedness (one-sided vs. two-sided), conclusiveness (explicit vs. implicit conclusions), and the gain- or loss framing approach. The versions were fundamentally adopted from a short essay about the WEF nexus on the website of UN-Water (n.d.). The following is a brief description of the 11 versions; for the original texts, please refer to the Appendix.

1. The Agenda Version (Title Only with No Message Body, One-Sided, Closed-Ended, Gain-Framed). The agenda version consisted of only a long title of "The number of media messages about the water-energy-food nexus, an approach to achieve water, energy, and food security simultaneously, has increased quickly since the approach was proposed in 2011" (sized like titles in other versions). It was the classical way of explicating the message agenda and excluded framing components. The title provided only positive-side information about the WEF-nexus concept (one-sided) with a conclusive (closed-ended) statement and framed the WEF nexus as achieving water, energy, and food security simultaneously (gain-framed). This version was randomly exposed to 26 (6.8%) subjects.

2. The Knowledge Version (One-Sided, Closed-Ended, Gain-Framed). The knowledge version contained a short title "The water, energy, and food security is in a nexus to achieve!" three informational paragraphs explaining how water, energy, and food were essentially linked and why they should be treated as a nexus, and a concluding paragraph repeating the title. The knowledge version provided a way of interpreting (framing) water, energy, and food issues as a nexus instead of being separate. Consistent with the agenda version for comparison, the knowledge version was also one-sided, closed-ended, and gain-framed. It was randomly exposed to 27 (7.1%) subjects.

3. The Attitudinal Version (One-Sided, Closed-Ended, Gain-Framed). The attitudinal version consisted of a short title "The nexus approach is a better concept to achieve water, energy, and food security!" and the three informational paragraphs, the attitudinal paragraph showing that the leading international organizations deemed the nexus approach as "urgent, important, innovative, and useful" for adoption, and a concluding paragraph repeating the title. It was also one-sided, closed-ended, and gain-framed, and was randomly exposed to 31 (8.1%) subjects.

4. The Behavioral Version (One-Sided, Closed-Ended, Gain-Framed). This behavioral version included a call-to-action title "Let's promote the nexus approach to achieve water, energy, and food security!" and the three informational paragraphs, the attitudinal paragraph, a paragraph calling for behavioral support for the WEF-nexus approach, and a concluding paragraph repeating the title. It was also one-sided, closed-ended, and gain-framed so that it could be compared with the first three versions to differentiate the persuasive effect of the agenda, knowledge, attitudinal, and behavioral components. It would also be used to compare with the following behavioral versions to test the persuasion effect of sidedness, conclusiveness, and gain- or loss-framing. This version was randomly exposed to 33 (8.6%) subjects.

5. Behavioral Version (One-Sided, Open-Ended, Gain-Framed). This behavioral version was created for comparison with the fourth version to test the persuasion effect of conclusiveness (closed-ended vs. open-ended). It looked the same (one-sided, gain-framed) as the fourth version except for using a question-style title and ending paragraph, both read: "Should we promote the nexus approach to achieve the water, energy, and food security?" The question-style title and ending paragraph featured open-endedness (an implicit conclusion or inconclusiveness). It was randomly exposed to 37 (9.7%) subjects.

6. Behavioral Version (Two-Sided, Closed-Ended, Gain-Framed). This behavioral version was created to be compared with the fourth version to test the persuasion effect of sidedness (one-sided vs. two-sided). It looked the same (closed-ended, gain-framed) as the fourth version except for adding a paragraph about the criticism of the WEF-nexus concept as "same old wine in new bottles," which was rebutted in the same paragraph. The title and the ending paragraph—"Despite the criticism, let's promote the nexus approach to achieve water, energy, and food security!"—both also mentioned the criticism, which made this version two-sided. This version was randomly exposed to 32 (8.4%) subjects.

7. Behavioral Version (Two-Sided, Open-Ended, Gain-Framed). This behavioral version, as a comparison with the sixth version, could further test the persuasive effect of conclusiveness (closed-ended vs. open-ended) because those two versions were both two-sided gain-framed. The inconclusiveness (open-ended or implicit nature) in this version of the message was featured again in the title and the ending paragraph, which both read "Given the criticism, should we promote the nexus approach to achieve the water, energy, and food security?" It could also be compared with the fifth version (one-sided, open-ended, gain-framed) to further test the effect of sidedness (one-sided vs. two-sided). It was randomly exposed to 31 (8.1%) subjects.

8. Behavioral Version (One-Sided, Closed-Ended, Loss-Frame). The eighth, also a behavioral version, was created for comparison with the fourth version to test the persuasion effect of the gain- or loss-frame. Both versions were one-sided closed-ended, but this version framed the WEF nexus as "to avoid the water, energy, and food crises" (loss-frame), compared with the gain-frame in the fourth version that framed the WEF nexus as "to achieve the water, energy, and food security." This version was randomly exposed to 32 (8.4%) subjects.

9. Behavioral Version (One-Sided, Open-Ended, Loss-Framed). The ninth version was created to compare with the fifth version (one-sided, open-ended, gain-framed) to further test the persuasion effect of the gain- or loss-frame and to be compared with the eighth version (one-sided, closed-ended, loss-framed) to further test the effect of the conclusiveness. This version was randomly exposed to 36 (9.4%) subjects.

10. Behavioral Version (Two-Sided, Closed-Ended, Loss-Framed). This tenth version was created for comparison with the sixth version (two-sided, closed-ended, gain-framed) to further test the effect of gain- or loss-framing and to be compared with the eighth version (one-sided, closed-ended, loss-framed) to further test the effect of sidedness. It was randomly exposed to 29 (7.6%) subjects.

11. Behavioral Version (Two-Sided, Open-Ended, Loss-Framed). The eleventh version was created to compare with the tenth version (two-sided, closed-ended, loss-framed) to further test the effect of conclusiveness, with the seventh version (two-sided, open-ended, gain-framed) to further test the effect of the gain- or loss-framing, and with the ninth version (one-sided, open-ended, loss-framed) to further test the effect of sidedness. It was randomly exposed to 36 (9.4%) subjects.

The blank page for the control group was randomly assigned to 32 (8.4%) subjects.

3.2 *The Manipulation-Check Variable: Answer Quality*

Following the message question, a question asked subjects whether the media message that they just read (1) only had a long title, (2) had knowledge components, (3) had attitudinal suggestions, (4) had behavioral suggestions, (5) just provided positive opinions about the WEF-nexus approach, (6) provided both positive and negative opinions about the WEF-nexus approach, (7) provided an explicit conclusion, (8) provided an implicit conclusion, (9) focused on what benefits the WEF-nexus approach could gain, (10) focused on what losses the WEF-nexus approach could avoid, or (11) the subjects did not read a message.

Among the subjects, 95 (35.7%) out of 266 subjects who were exposed to a version with a behavioral component answered the manipulation check question correctly. The correct rates for other groups: 94 (31.6%) out of 297 for subjects exposed to a version with an attitudinal component (including the attitudinal version, and all behavioral versions); 177 (54.6%) out of 324 for subjects exposed to versions with an information component (including the knowledge version, attitudinal version, and all behavioral versions); 5 (19.2%) out of 26 for subjects exposed to the agenda version; 74 (57.8%) out of 128 for subjects exposed to a two-sided version; 82 (36.9%) out of 222 for subjects exposed to a one-sided version; 64 (30.5%) out of 210 for subjects exposed to a closed-ended version; 39 (27.9%) out of 140 for subjects exposed to an open-ended version; 102 (47%) out of 217 for subjects exposed to a gain-framed version; 39 (29.3%) out of 133 for subjects exposed to a loss-framed version; 24 (75%) out of 32 for subjects exposed to no version (in the control group).

Although the manipulation-check questions were not answered optimally, the physical characteristics of the message stimulus for levels and factors were present. In message effects studies, O'Keefe (2003) argues, when message variations are defined by intrinsic message features, manipulation checks are unnecessary. He suggests examining the possible mediation function of the manipulation check questions as a measure of subjects' psychological state in message effects studies. Following this suggestion, a control variable of "answer quality" was created. Respondents earned one point if answering one manipulation question correctly ($N = 402$, $Min = 0$, $Max = 6$, $M = 1.98$, $SD = 1.62$).

3.3 Response Variables (Asked After the Message Stimulus)

WEF-Nexus Agenda. The salience or importance of the issue in subjects' minds was measured by how, on a scale of 1 = strongly disagree to 7 = strongly agree, subjects evaluated the statement that "how to manage water, energy, and food security is an important topic to discuss" (ranging from 1 to 7, $N = 365$, $M = 5.63$, $SD = 1.20$), a classical measurement of public agenda (McCombs & Reynolds, 2009).

Subjective WEF-Nexus Knowledge. After the message stimulus, subjects were asked to rate on a Likert scale (strongly disagree = 1, neutral = 3, and strongly agree = 5) about the statement "I consider myself well informed about issues in the water-energy-food nexus" ($N = 366$, $M = 3.54$, $SD = 1.51$).

Objective WEF-Nexus Knowledge. Five true/false questions based on the informational component of the message were used to measure subjects' recall of the information in the message. Subjects earned one point for each correct answer. The index thus ranged from 0 to 5 ($N = 402$, Cronbach's $\alpha = .6$, $M = 2.24$, $SD = 1.41$).

WEF-Nexus Frame. Subjects were also asked to mark, on a scale of 1 = strongly disagree to 7 = strongly agree, how much they agreed with the statement, "The water, energy, and food security is in a nexus," a way of interpretation that messages with informational components aim to establish. The statement was a measure of the subjects' frame (cognitive recognition) of the issue of the WEF nexus (ranging from 2 to 7, $N = 367$, $M = 4.91$, $SD = 1.15$).

WEF-Nexus Attitude. The same index of six questions used in the first round was used in the second round, but on a scale of 1 = strongly disagree to 7 = strongly agree ($N = 357$, Cronbach's $\alpha = 0.84$, $M = 30.53$, $SD = 5.37$, $Min = 15$, $Max = 42$).

WEF-Nexus Behavioral Intention. The same index of the five items used in the first round was used in the second round on a scale of 1 = strongly disagree to 7 = strongly agree (Cronbach's $\alpha = 0.90$, $N = 364$, Min = 5, Max = 35, $M = 20.87$, $SD = 5.48$).

Environmental Behavioral Intention. The same nine items used in the first round were used again in the second round, also on a scale of 1 = strongly disagree to 7 = strongly agree ($N = 358$, Cronbach's $\alpha = 0.89$, $M = 47.97$, $SD = 13.20$).

To better understand the persuasion effectiveness, the following variables were also measured and controlled, together with demographic and socioeconomic variables. The variable of answer quality created from the manipulation variables were also controlled in some analyses.

3.4 *Control Variables (Asked Before the Message Stimulus)*

Because some subjects participated in the first round of the survey and may have learned something about the WEF nexus, and because the survey experiment was administered in a real-life setting and lasted for five months to gather enough responses, the following variables were necessary to be controlled in the analyses.

Subjective WEF-Nexus Knowledge 1. Subjects were asked how much they knew about the WEF nexus, with 1 = never heard to 10 = very knowledgeable (M = 2.71, SD = 2.11), which served as a control variable in some analyses.

Keeping Up with Science News. On a scale from 1 = not at all to 10 = a lot, respondents were asked to mark how much they kept up with news about science (M = 5.72, SD = 2.30).

Religiosity. The variable was again measured by the sum of standardized values of the question on the frequency of attending religious services aside from weddings and funerals (N = 363, Min = 1, Max = 6, M = 2.96, SD = 1.58) and the importance of God(s) in life (N = 321, Min = 0, Max = 100, M = 58.28, SD = 40.08). This scale was reliable (Cronbach's α = 0.81, N = 319, Min = −2.7, Max = 2.97, M = 0.07, SD = 1.83).

Science Literacy. The nine-item science literacy test used in the first round (National Science Board, 2016) was used in the second round again, which again ran from 0 to 9 (Cronbach's α = 0.6, N = 402, M = 7.26, SD = 1.65).

Demographic, socioeconomic, and some other variables were also measured and controlled in some analyses to test the hypothesized effects of the message levels and factors on the audience of American college students in the following chapters.

References

Ang, L., & Eisend, M. (2018). Single versus multiple measurement of attitudes: A meta-analysis of advertising studies validates the single-item measure approach. *Journal of Advertising Research, 58*(2), 218–227. https://doi.org/10.2501/JAR-2017-001

Annas, G. J. (2006). Intelligent judging – Evolution in the classroom and the courtroom. *The New England Journal of Medicine, 354*(21), 2277–2281. https://doi.org/10.1056/NEJMlim055660

Carter, J. (2005). *Our endangered values: America's moral crisis.* Simon & Schuster.

Eccles, J. C. (1989). *Evolution of the brain: Creation of the self.* Routledge.

Fischer, F. (2000). *Citizens, experts, and the environment: The politics of local knowledge.* Duke University Press.

Gastil, J., & Xenos, M. (2010). Of attitude and engagement: Clarifying the reciprocal relationship between civic attitudes and political participation. *Journal of Communication, 60*(2), 318–343. https://doi.org/10.1111/j.1460-2466.2010.01484.x

Han, T.-I. (2019). Objective knowledge, subjective knowledge, and prior experience of organic cotton apparel. *Fashion and Textiles, 6*(4). https://doi.org/10.1186/s40691-018-0168-7

Hawking, S. (1988). *A brief history of time: From the Big Bang to black holes.* Bantam Books.

Hesaraki, M., Akbarizadeh, M., Ahmadidarrehsima, S., Moghadam, M. P., & Izadpanah, F. (2020). Knowledge, attitude, practice and clinical recommendations of health care workers towards COVID-19: A systematic review. *Reviews on Environmental Health, 36*(3), 345–357. https://doi.org/10.1515/reveh-2020-0099

Kanu, S., James, P. B., Bah, A. J., Kabba, J. A., Kamara, M. S., Williams, C. E. E., & Kanu, J. S. (2021). Healthcare workers' knowledge, attitude, practice and perceived health facility preparedness regarding COVID-19 in Sierra Leone. *Journal of Multidisciplinary Healthcare, 14*, 67–80. https://doi.org/10.2147/JMDH.S287156

Kruglinski, S. (2007). Islam's scientific legacy. *Discover, 28*(7), 40.

Lee, I. A., & Preacher, K. (2013, September). Calculation for the test of the difference between two dependent correlations with no variable in common [Computer software]. *Quantpsy.org.* Retrieved March 4, 2025, from http://quantpsy.org/corrtest/corrtest2.htm

Lee, O. (1998). Diversity and equity for Asian American students in science education. *Science Education, 81*(1), 107–122. https://doi.org/10.1002/(SICI)1098-237X(199701)81:1<107::AID-SCE6>3.0.CO;2-M

Lee, S., Diehl, T., & Valenzuela, S. (2022). Rethinking the virtuous circle hypothesis on social media: Subjective versus objective knowledge and political participation. *Human Communication Research, 48*(1), 57–87. https://doi.org/10.1093/hcr/hqab014

McCombs, M., & Reynolds, A. (2009). News influence on our pictures of the world. In J. Bryant & D. Zillmann (Eds.), *Media effects: Advances in theory and research* (3rd ed., pp. 1–16). Routledge.

Nasr, S. H. (n.d.). Islam and modern science. *MIT Website*. Retrieved March 4, 2025, from https://web.mit.edu/activities/mitmsa/NewSite/libstuff/nasr/nasrspeech1.html

National Science Board. (2016). Science & engineering indicators. *National Science Foundation Website*. Retrieved March 4, 2025, from https://www.nsf.gov/nsb/publications/2016/nsb20161.pdf

O'Keefe, D. (2003). Message properties, mediating states, and manipulation checks: Claims, evidence, and data analysis in experimental persuasive message effects research. *Communication Theory, 13*(3), 251–274. https://doi.org/10.1111/j.1468-2885.2003.tb00292.x

Riegle-Crumb, C., & King, B. (2010). Questioning a white male advantage in STEM: Examining disparities in college major by gender and race/ethnicity. *Educational Researcher, 39*(9), 656–664. https://doi.org/10.3102/0013189X10391657

Russell, B. (1961). *Religion and science*. Oxford University Press.

Sideris, L. H. (2017). *Consecrating science: Wonder, knowledge, and the natural world*. University of California Press.

UN Water. (n.d.). Water, food and energy. *United Nations Website*. Retrieved March 4, 2025, from http://www.unwater.org/water-facts/water-food-and-energy/

Yao, Q., Chang, C.-F., Joshi, P., & McDonald, C. (2024). Climate change versus the water-energy-food nexus: The oldness or newness of the scientific issues as a factor in the deficit model and the hierarchy of response model. *Environment, Development and Sustainability, 26*, 823–840. https://doi.org/10.1007/s10668-022-02735-3

Yao, Q., Joshi, P., Chang, C.-F., McDonalds, C., Tran, J., Wheeler, W., & Hou, S. (2018). Advocating a new approach to governing water, energy, and food security: Testing the effects of message inoculation and conclusion explicitness in the case of the WEF nexus. *Journal of International Crisis and Risk Communication Research, 1*(1), 111–133. https://jicrcr.com/index.php/jicrcr/article/view/6/6

Yao, Q., Martin, M. C., Yang, H.-Y., & Robson, S. (2019). Does diversity hurt students' feeling of oneness? A study of the relationships among social trust, university internal brand identification, and brand citizenship behaviors on diversifying university campuses. *Journal of Marketing for Higher Education, 29*(2), 209–229. https://doi.org/10.1080/08841241.2019.1638482

CHAPTER 5

The Five-Level Model of Message Effects: Promoting the WEF Nexus at Different Levels

To understand the persuasiveness of the messages, researchers need to understand how the audience processes and responds to these messages. The hypodermic needle or magic bullet theory is probably the first theory of persuasion (particularly through mass media) on audience responses, which assumes that the audience will respond with according actions immediately after receiving the persuasion. The theory emerged at the beginning of the last century when scholars started to be interested in communication effects. Although it has laid a solid foundation for later research and theorization of persuasion and media effects, it has almost turned into a farce because it assumes the audience as a totally passive receiver and ignores the differences and moderating roles of the audience characteristics and information processing procedures (Yao, 2008). Later researchers, frustrated in finding the communication effects predicted in the hypodermic or magic bullet theory, realized that analyzing and understanding the audience is key to mastering the nature, process, and effects of communication. Understanding how the audience processes messages is critical for communication researchers to construct effective messages. Although this body of research has largely been conducted in the context of mass communications, findings can also be applied to strategic communications, which are specified with a behavioral-changing goal among the target audience through specially designed messages (Besley & Dudo, 2022; French & Russell-Bennett, 2015; McKenzie-Mohr et al., 2012).

Q. J. Yao, *Marketing the Water-Energy-Food (WEF) Nexus*,
https://doi.org/10.1007/978-3-031-93148-2_5

1 Communication Effects and a Definition

Communication scholars have long been exploring whether messages, particularly those conveyed through mass media, have effects on the audience, and if so, what those effects are. The answer is critical due to its significant impact on social, political, and economic matters. For instance, although many parents believe that violent video games would teach their kids to kill, and some states banned selling violent video games to minors, the courts often struck down those bans, citing the First Amendment protection of video games as speeches and social scientific research not proving video games' physiological or psychological harm to minors (Pember & Calvert, 2008).

The path to identifying communication effects has been frustratingly nonlinear throughout the past century. The hypodermic needle or magic-bullet model was totally overthrown in the 1940s when Lazarsfeld and his team tested the effects of the ads on the presidential elections of 1940 and 1944 (Lazarsfeld et al., 1944; Lazarsfeld & Merton, 1948), in which they found voters influenced by not campaign persuasion but the people they knew. The limited-effect theory was then established and dominated communication studies for decades (Chaffee & Lieberman, 2001), until the 1970s when strong-effect models such as agenda-setting (McCombs & Shaw, 1972), priming (Iyengar, 1991; Iyengar & Kinder, 1987), and framing (Gamson & Modigliani, 1989; Kahneman & Tversky, 1984; Tversky & Kahneman, 1981) emerged again.

Despite the volume of research supporting the existence of message effects on the audience, communication researchers still face challenges from sociologists, who often attribute communication effects to real-world events (Gans, 1993; Tuchman, 1978, 1993). To address their challenges, a clear working definition of communication effects is needed.

1.1 Issues with the Limited-Effect Theory

Klapper's (1960) synthesized review of thousands of effects studies has been taken as an authoritative presentation of the limited-effects theory (Katz, 2001). Mass communication, concludes Klapper (1960), generally produces no effect on the audience; it mostly reinforces the existing opinion, sometimes leads to minor attitude changes, and rarely causes conversion. Mass communication effects are moderated by its characteristics such as textual organization, source and media credibility, and the existing climate of public opinion. It functions through these and other moderating or mediating factors and only produces direct effects when those factors are inoperative or impelling toward change (Klapper, 1960).

Based on Klapper (1960), Lazarsfeld's team found little direct effect of mass communication mainly because they adopted an understanding of the communication effects that was narrow (Katz, 2001) in two ways:

First, unclear definition of mass communication. Direct effects of mass communication are usually easy to find in laboratories, which can control confounding factors so that only experiment stimuli function. But even in those labs, interventions such as the left-over effects, like audience predisposition, still exist if the issues in the stimuli are well published and controversial. Lazarsfeld's studies, however, were conducted in natural settings, and they were about "opinions on controversial issues" (Chaffee et al., 1970, p. 648). The audiences could have easily been exposed to all information about those issues from the media ecosystem, which eliminated the advertising effects that Lazarsfeld planned to observe (Klapper, 1960). So, despite the dominance of the limited-effect theory in the 1950s–60s, no political candidate dared to stop campaigning through mass media: when the persuasion from one side stopped, effects from the other side would emerge.

Second, unclear definition of effects. Lazarsfeld just focused on mass communication's ability to change voting intentions (McCombs & Reynolds, 2009), missing the effects of other levels and directions. Communication effects, however, most likely happen at the levels of the acquisition of political knowledge or the building of political interests (Chaffee et al., 1970). Also, the same information can form an opinion in one part of the audience but a directly contrary one in the other, and an individual can have several contradictory opinions, the "considerations" (Zaller, 1992, p. 118), toward an issue simultaneously. When a message increases a weaker consideration to match the contrary dominating considerations in strength, the individual cannot provide an opinion, falling into ambivalence. More communication may drive the person out of the ambivalence for another opinion (Zaller, 1992). Even if the person sticks with the original opinion or ambivalence, the considerations' strength might have changed. Aggregated audience data on opinion direction cancel out these important individual changes and make it appear that "nothing" has happened (Chaffee et al., 1970, p. 666).

In the Lazarsfeld 1940 study, several months of campaign persuasion reinforced the original pre-campaign intention of 53% of the 600-panel members, changed 26% from decided to undecided or from undecided to decided, and converted 5% to change partisanship. In their similar 1948 study, the one-month campaign reinforced 66% of the 760-respondent panel's partisanship, switched 17% from a party to neutral, and converted 8% to different parties (Klapper, 1960). But Lazarsfeld and associates just "look[ed] for media influences in one attitudinal direction, rather than ask[ed] the more fundamental question of whether the person forms any opinion" (Chaffee et al., 1970, p. 659). They should have counted those offset effects or the reinforcement when the message effect was consistent with the audience's predisposition (Katz, 2001; Potter, 2012). Late research upholds communication effects measured as much more than just behavior changes (McQuail, 2010; Potter, 2012) and at different levels (Yao et al., 2020) and directions. More subtle or segmented audience analysis can examine message effects more comprehensively and precisely.

These two lessons encourage later researchers to focus more on other levels of effects, such as agenda (i.e., issue importance. McCombs & Shaw, 1972), knowledge (Chaffee et al., 1970), evaluation criteria (Iyengar, 1991; Iyengar & Kinder, 1987), and frames (Gamson & Modigliani, 1989; Kahneman & Tversky, 1984; Yao et al., 2020). Those studies, together with the concept of information ecology, enable us to develop a clear explication of communication effects.

1.2 A Definition of Communication Effects

Three questions need to be clearly answered to propose a definition of communication effects: First, what are considered communication effects? McQuail (2010) defines one type of them, media effects, as "simply the consequences of what the mass media do, whether intended or not" (p. 463) and provides a comprehensive diagram to chart media effects along the dimensions of time and levels of information process. Similarly, Potter (2012) defines media effects as "those things that occur as a result—either in part or in whole—from media influence" (p. 38). However, many of the potential media effects are too complex or subtle to be measured by the social scientific techniques available (McQuail, 2010). Communication researchers are still less interested in understanding the differences between short-term and long-term effects or even effects in different directions, but they have learned to pay attention to effects at different levels of information processing: attention, awareness, knowledge, attitudes, and behaviors (Yao, 2008; Yao et al., 2020). Warning that early communication effects research devoted itself too much to attitudinal shifting but too less to cognitive change, McLeod and Reeves (1980) hypothesize the existence of a strong causal flow from message exposure to knowledge gain, attitude shift, and behavior change. McGuire (1989) also proposed a hierarchical model that classifies campaign effects into 12 levels, ranging from exposure to post-behavioral consolidation, which he later condensed to six levels (McGuire, 2013). Changes and reinforcement at any of those levels are communication effects, and changes across those levels also occur along McQuail's (2010) time axis.

Second, what is considered communication? To avoid Lazarsfeld's error, in the definition, "communication" should be the entire body of information sources as one information ecosystem to which the audience is exposed, which could be analyzed respectively. The omission of any information source leads to inaccurate calculation of the effects. To simulate such a media ecosystem, McCombs and Shaw (1972) content analyzed five newspapers, two news magazines, and two network news programs, sources that provided "almost all the mass media political information" to the community of their audience (p. 178). The concept of a communication ecosystem is also crucial to understanding why media in open societies, where media content can be conflicting, normally have less impact than their counterparts in authoritarian societies, where media content is normally coordinated and consistent.

Third, can researchers distinguish communication effects from the effects of real-world indicators (Gans, 1993; Kosicki, 1993)? If communication is understood as an information ecosystem, the real-world indicator has little role in the definition of communication effects. In a mass society, only through mass or interpersonal communication can individuals interact with the world beyond their personal experiences (see Fig. 5.1). Communication is the principal conduit between the world beyond personal experiences and the images of that world people hold in their minds, the pseudo-environment (Lippmann, 1922/1997) or living environment (Mutz, 1998). The real-world events beyond people's personal experiences (such as Event A in Fig. 5.1) have little influence on people until those events are communicated, as evidenced in Lippmann's (1922/1997) Island in 1914, where the Englishmen and Frenchmen were still friends with the Germans six weeks after the outbreak of the WWI until they learned the news of the war from a ship visiting the island. Indeed, what effects can the real-world indicator of a tree falling in the forest have on us if we do not see it and nobody communicates it to us?

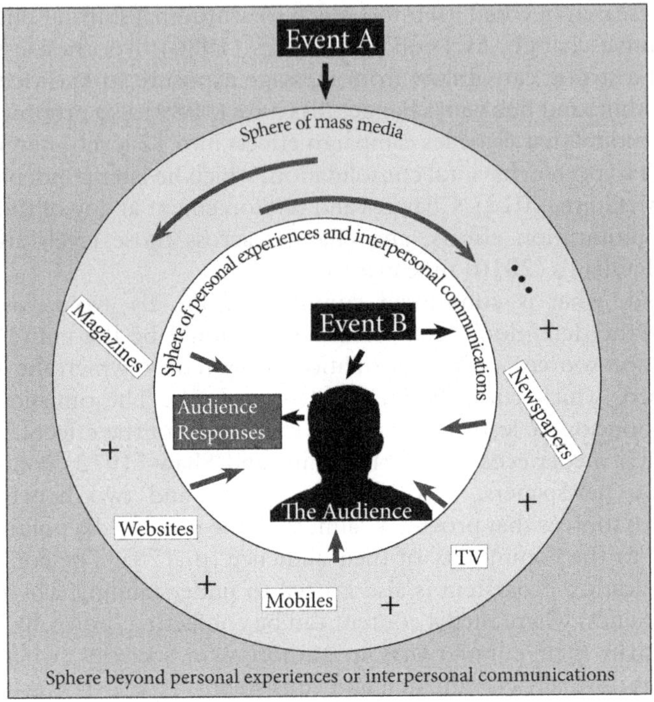

Fig. 5.1 Communication effects as audience responses to the communication ecosystem. (Notes: Event B is within the audience's sphere of personal experiences or interpersonal communications and can directly affect the audience. Event A, in contrast, is beyond the audience's sphere of personal experiences or interpersonal communications and can only impact the audience through the mass communication ecosystem)

Research has also consistently shown that communication, rather than real-world indicators (issues or events), leads to changes in public perception of these issues or events (Rogers et al., 1993). For example, Danielian and Reese (1989) find that the issue of drug abuse received the most media coverage and thus stayed high on the U.S. agenda from 1986 to 1989, while the long-term trend of the number of drug-related deaths in the U.S. had slowly and linearly declined during that period. Gans (1979) notes that child abuse and wife beating existed without social notice until journalists created the news category of *Family* and began to report them, creating a social discussion addressing those issues. Fishman (1980) also finds that when a major "crime wave" hit the media in New York City, the statistics of the crime had actually already decreased. Nevertheless, the intensive media-produced discussion of the "crime wave" made the mayor, the state legislature, and the local communities act to respond and spread a nationwide fear of this particular type of crime.

Communication effects, therefore, should be explicated *as all responses from the audience as a result of being exposed to the communication ecosystem, that is, the entire information environment.* Such communication effects are usually named persuasiveness in the context of strategic communications, particularly when they are aligned with the persuasion purpose. Communication effects can be analyzed with different units (individual, group, national, etc.; Katz, 2001; McLeod et al., 1995), at different levels (cognitive, affective, or conative. McGuire, 2013; Potter, 2012), within different time frames (immediate response, short-term, long term, or institutional, etc.; Katz, 2001;McQuail, 2010 ; Potter, 2012), and as planned or unplanned (McQuail, 2010; Potter, 2012). Developing delicate and precise measurements is essential for capturing communication effects in different directions that may have been canceled out in previous studies.

Although communication effects can be analyzed at different dimensions, in the following sections of this chapter, we will focus on the multiple levels of communication effects, which have accumulated a large body of research but still need further articulation to deepen our understanding.

2 The Hierarchy of Effects and the Deficit Model

Some communication researchers and practitioners have long noticed the importance of understanding the audience's psyche, which is why psychologists were the first group of social scientists entering the field of audience research, especially consumer psychology. One of their findings, the AIDA model, is still influential in the industry of strategic communication (Hassan et al., 2015), which includes four levels of audience responses: attention, interest, desire, and action. As the time and volume of persuasion increase, audience responses increase along the levels. It remains a foundation for understanding later models of audience psychology.

2.1 *The Hierarchy of Response Models*

More elaborative than the AIDA model, Lavidge and Steiner's (1961) widely circulated hierarchical responses model defines persuasion effects as six stages: awareness, knowledge, liking, preference, conviction, and purchase. Proposed in the same year, the model of defining advertising goals for measured advertising results (DAGMAR) defines four stages of persuasion effects: awareness, comprehension, conviction, and action, holding that any persuasion campaign plan must identify specific, measurable goals in one or multiple stages, target at a certain segment of the population, and establish a completion time (Barry, 2002; Weilbacher, 2001). The next year, the well-known "diffusion of innovation" theory (Rogers, 1962, 2003) was proposed, arguing that innovative ideas normally go through six steps in their audience's mind: knowledge, persuasion, decision, implementation, and confirmation. Then came William McGuire's (1989, 2013) early 12- and later six-level model of campaign effects (message presentation, attention, comprehension, yielding/conviction, retention, and behavior).

Also hierarchical is the consumer-based brand equity (CBBE) model (Keller, 2012), which holds that the power of the brand exists entirely in people's feelings and experience and is built through several stages resulting from the accumulation of brand persuasion and user experience. In the first stage, brand salience, brands get basically understood. Then brand performance and brand imagery are established. Brand performance is the comprehensive feeling of the brand's ingredients, performance, reliability, style, price, and so on. Based on it, brand image is formed as the perception of the brand's overall user image, purchasing and using environments, intrinsic value, and history. Continuing persuasion and experience about the brand help people to form brand judgments, the overall assessment of the quality, credibility, and excellence of the brand, and brand feelings, perceiving the brand as warm, cute, safe, exciting, and so on. Entering the final stage, brand resonance, consumers turn to be loyal to and emotionally dependent on the brand and want to join the community formed by the brand users.

Other hierarchy of response models include the oft-cited theory of reasoned action (TRA), theory of planned behavior (TPB), and model of goal-directed behaviors (MGB), all explaining how desire turns into action. In these models, the audience's subjective norms (perception of the social pressure of taking or not taking the action), perceived behavioral control (evaluation of the self's capacity of taking the action), attitude (personal favorability to take the action), and anticipated emotions after taking the action all moderate whether the audience desires turn into action (Lee et al., 2018).

2.2 *The Cognitive-Affective-Conative Models*

The hierarchy of response models discussed above, albeit similar in nature, consists of a variety of levels. They later led to the popularity of a condensed three-step model, each containing some small steps: cognitive (attention, exposure, awareness, recognition, comprehension, and recall), affective (attitude change and involvement), and conative (intention to act and behavior. Leckenby & Wedding, 1982; Ray, 1973). The three-level model is so popular that even natural scientists also examine the effects of information processing in their fields at the levels of knowledge, attitude, and behavior (or practice; Xu et al., 2010). The deficit model, the first dominant theory of science communication proposed in the 1980s, is a part of this cognitive-affective-conative model (Belch & Belch, 2014; Kotler & Keller, 2006; McGuire, 1989), predicting that the increase of information provision of science increases public scientific knowledge, which increases public support for scientific affairs (Miller, 2001). It is still influential in the fields of science as well as environmental and health communications (Bettinghaus, 1986; Hart & Nisbet, 2012; Scheufele, 2013). When the nature of the issues is complex, confounders such as ideological, cultural, or economic biases interfere, and the knowledge-attitude linkage often turns illusive (Belch & Belch, 2014; Kollmuss & Agyeman, 2002; Scheufele, 2013), which indicates the needs for further audience elaborative analyses.

Ray (1973) elaborates on the audience characteristics and holds that the audience might experience these three steps in different chronological orders when processing different types of persuasive messages. The regular cognitive-affective-conative model is the standard learning model, which goes from knowledge to attitude to behavior. In some cases, nevertheless, the audience may also be forced by the persuasive messages and other pressures to behave first and then feel and learn about the topic in the messages, forming a dissonance-attribution model. When dealing with a topic that does not require much effort to act on, the audience may also use a low-involvement model: acting after learning the messages, and then forming an attitude, going through the order of cognition, conation, and affection (Ray, 1973).

Ray's (1973) three models are further developed into the FCB Grid (Vaughn, 1986). This grid, based on the relationship between products/services (issue or topic in the message) and consumers (audience), classifies consumers into four quadrants along two axes: high- and low-involved, and think-feel. Consumers in quadrant one are highly involved and use the thinking strategy to decide whether to buy, such as when consumers buy cars, houses, or insurance. These "thinkers" use the learning-feeling-behavior processing model; so, the persuasion should be informative. Consumers in quadrant two are equally highly involved but prefer a feeling strategy, hence called "feelers," such as those who buy jewelry, cosmetics, and fashion products/services. "Feelers" use the feeling-learning-behavior model, and the persuasion should emphasize tastes. Quadrant-three consumers are low-involvement and use the thinking strategy. So, they are called "actors," such as buyers of food and furniture. They process the persuasion in an order of behavior-learning-feeling, and the persuasion can use a small but frequent strategy to remind them to purchase. Quadrant-four consumers, "responders," are low-involved and focus on emotional decisions, such as buyers of cigarettes, wine, and candy. Their order to process the persuasion is behavior-feeling-learning; so, the persuasion strategy should capture consumers' attention (Vaughn, 1986). Many copywriters used this grid to guide their creation of persuasive messages.

3 Multi-levels of Message Stimuli and Audience Responses

In decades of exploring communication effects, communication scholars have not studied audience responses and message stimuli at multiple levels, but separated in agenda-setting, priming, and framing studies (Scheufele & Tewksbury, 2007). These different lines of research create confusion in the systematic understanding of communication effects (Scheufele & Iyengar, 2014; Scheufele & Tewksbury, 2007; Weaver, 2007); so, attempts to integrate them have been made (McCombs & Reynolds, 2009; Tewksbury & Scheufele, 2009;Yao, 2008 ; Yao et al., 2020). Among these three lines of research, agenda-setting is the first to emerge and challenge the dominance of the limited-effect theory in the late 1960s (Katz, 2001; Rogers et al., 1993).

3.1 The Agenda-Setting Research

Agenda-setting research's popularity in the 1970s can partly be attributed to researchers' dissatisfaction with the dominant limited-effect theory (Kosicki, 1993), which needed to be replaced by an "alternative to the scholarly search for direct media effects on attitude change and overt behavioral change" (Rogers et al., 1993, p. 73). Decades of agenda-setting research have found evidence supporting a robust positive correlation between the salience of an issue, topic, or aspect in media (media agenda) and its salience in the audience's mind (public agenda. Yao et al., 2020), which "seemed to override selective exposure, perception, and retention—forces that contribute to reinforcement rather than a change of attitudes and opinions in earlier studies of media effects" (McCombs et al., 1991, p. 14).

The best way to clarify empirical theories is to analyze their conceptualization and operationalization of their variables (Yao, 2008; Yao et al., 2020). In agenda-setting research, agendas are conceptualized as salience (Kosicki, 1993; McCombs & Shaw, 1993), or, "the importance that media and audience accord to an event" (Edelstein, 1993, p. 85). Importance is operationalized as the hierarchy, or rank, of the issues (Sheafer & Weimann, 2005). More specifically, media or message agenda is measured by the amount of space or time devoted to particular issues in the news or messages; public agenda is measured either by the amount of attention that people pay to the issues or the degree of importance that people attribute to the issues (Iyengar & Kinder, 1987; Kosicki, 1993). Agenda-setting researchers traditionally use a specific research design that combines mass media content analysis and public opinion surveys (Rogers et al., 1993), in one of the four ways discussed in the "Acapulco typology" (McCombs & Reynolds, 2009). The theory basically holds that "[i]n any event, issues prominent on the news agenda are perceived by the public to be important, and, over time, frequently become the priority issues on the public agenda" (McCombs et al., 1991, 7. See also McCombs & Reynolds, 2009; McCombs & Shaw, 1972). The same conceptualization and operationalization apply to the second level of agenda-setting that emerged later, in which the unit of analysis moves from issues to attributes, the "specific aspects" of the issues or objects (McCombs & Reynolds, 2009; McCombs & Shaw, 1993). Agenda-setting has been criticized as bypassing the real persuasion and the cognitive structure of the issues (Moy et al., 2016; Moy & Bosch, 2013); calls for higher orders of cognitive analysis have almost always accompanied agenda-setting research (Edelstein, 1993; Yao et al., 2020).

3.2 *The Priming Research*

Priming research has been conducted in two groups: political scientists (Iyengar & Kinder, 1987), and psychologists (Roskos-Ewoldsen et al., 2002). Political scientists' priming (Iyengar & Kinder, 1987) is a temporal extension of agenda-setting (McCombs & Reynolds, 2009; Scheufele & Tewksbury, 2007; Sheafer & Weimann, 2005), holding that the media agenda primes an issue as a criterion for the public to evaluate a politician's overall performance (Miller & Krosnick, 2000; Moy & Bosch, 2013). The independent variable is media exposure or agenda, and the dependent variable is how much the participants weigh the viewed issue in their overall evaluation of the politicians, mainly the presidents. Iyengar and Kinder (1987) argue that priming effects happen because "problems covered by television news become more accessible and therefore more important in the viewer's political calculus" (70). Priming implies that media exposure influences audience attitudes, even though attitude is not the independent variable in this research.

Following the rationale of the first level of agenda-setting research, Iyengar and Kinder (1987) also avoid exploring the persuasion function of the media messages, because "political persuasion is difficult to achieve, but agenda-setting and priming are apparently pervasive" (p. 117). They, therefore, stated that "the power of television news—and mass communication in general appears to rest not on persuasion but on commanding the public's attention (agenda-setting) and defining criteria underlying the public's judgment (priming)" (p. 117). Later, Sheafer and Weimann (2005) extend the dependent variable of the priming research to electoral behaviors based on two reasons: (1) previous research suggests a high positive correlation between the evaluation of the presidents and voting for or against them; and (2) when issues of a political party are emphasized that party will be inevitably helped in elections.

Psychologists study priming effects from a different approach and use behaviors or behavioral intention as both the independent and the dependent variables (Roskos-Ewoldsen et al., 2002). A number of studies have shown that viewing violent television programs increases aggression intentions and behaviors and that exposure to music videos containing stereotypical images of men and women leads people to have more of such stereotypical impressions and behave accordingly (Moy & Bosch, 2013). A synthetic review, therefore, defines psychological priming as "the effects of the content of the media on people's later behavior or judgments related to the content" (Roskos-Ewoldsen et al., 2002, p. 97).

3.3 *The Framing Effects Research*

Framing effects research originated from the prospect theory in the economic and psychological literature (Kahneman & Tversky, 1984; Tversky & Kahneman, 1981) and was then extended to political science (Druckman, 2001; Entman, 1993; Iyengar, 1991) and communication (Cacciatore et al., 2016; O'Keefe & Jensen, 2006; Scheufele, 1999, 2000; Yao, 2007) research. The term "framing" has different meanings in different literatures; overall, it is the process of messages' frames influencing audiences' frames of a certain issue, which is also called framing effects or frame-setting (Scheufele, 1999, 2000; Yao et al., 2020). Message frames are aspects of the topic that are chosen by the communicators to imply "potential problem definitions, explanations, evaluations, and recommendations" (Entman, 1993, p. 54).

Message frames have been operationalized as gain- or loss-aspects (O'Keefe & Jensen, 2006; Tversky & Kahneman, 1981), thematic (presenting collective or general evidence, usually in an in-depth and interpretative way and leading the audience to attribute responsibility to the government or society) or episodic (depicting concrete events, usually in a visually striking way and leading the audience to attribute responsibility to the individuals) approaches (Iyengar, 1991), or "logically equivalent (but not transparently equivalent) statements of a problem (that) lead decision makers to choose different options" (Druckman, 2004, 1024), which influence public opinion toward the problem (Druckman & Nelson, 2003), such as framing a Ku Klux Klan rally as an issue of free speech or public safety (Druckman, 2004).

Communication researchers have found that different frames of the same issue affect the audience's decision-making regarding university funding reduction (Price et al., 1997); issue perception, and consequent attitudes toward stem cells, the environment (Shen, 2004), and welfare (Shen & Edwards, 2005); and personal framing of religion (Yao, 2007). Some studies, nevertheless, fail to find a relationship between media frames and frames in the public's mind (Edy & Meirick, 2007; Yao et al., 2020).

3.4 *The Debate Between Agenda-Setting and Framing Effects Researchers*

The debate about the difference between agenda-setting research, particularly at the second level that focuses on the "attributes" instead of "issues" or "people," and framing effect research has lingered for decades (Yao et al., 2020), which is mainly caused by confusing conceptualizations of media frames and agendas (Cacciatore et al., 2016; McCombs & Reynolds, 2009; Scheufele & Iyengar, 2014; Scheufele & Tewksbury, 2007; Tewksbury & Scheufele, 2009). Examining their conceptualization and operationalization of the independent and dependent variables can help clarify the confusion and lay down the foundation of a multi-level model of communication effects.

In the debate, agenda-setting researchers hold that "both the selection of objects for attention and the selection of frames for thinking about these objects are powerful agenda-setting roles" (McCombs & Shaw, 1993, p. 62) because the frame is a special attribute that reflects the dominant perspective on the object (McCombs & Reynolds, 2009). Even if second-level agenda-setting and frame setting are not identical, they should be very similar (Weaver et al., 2004). Framing researchers, nevertheless, argue that framing effects go beyond agenda-setting to the sphere of meaning construction and cognitive analysis (Pan & Kosicki, 1993), with the devices of metaphors, exemplars, catchphrases, depictions, and visual images (Borah, 2011; Gamson & Modigliani, 1989; Pan & Kosicki, 1993). The argument of the framing researchers emphasizes that, at least, the dependent variables of those two lines of research are measured in different ways (Cacciatore et al., 2016; Edy & Meirick, 2007; Scheufele, 2000; Scheufele & Iyengar, 2014; Scheufele & Tewksbury, 2007; Tewksbury & Scheufele, 2009), although a small portion of communication scholars like to use framing and second-level agenda-setting interchangeably (Borah, 2011). To further clarify second-level agenda-setting and framing effects and make empirical operationalization convenient, a group of framing researchers drastically limit the intention and extension of framing to only equivalent framing (Cacciatore et al., 2016; Druckman, 2004).

Indeed, second-level agenda-setting and framing effects studies have different dependent variables: public (audience) agenda, the dependent variable of agenda-setting, refers to the rank or perceived importance of issues or attributes; audience framing, the dependent variable of framing effect research, usually is how audience members think of the issue and what will be their attitude and behavioral intention toward the issue. Beyond the agenda level, agenda-setting researchers are aware that "it is likely that increased salience of an issue will result in more public knowledge and stronger public opinions, but it is less certain what direction that opinion will take" (McCombs et al., 1991, p. 19). Framing, however, posits that how the messages interpret an issue will influence the audience's interpretation (knowledge) (Edy & Meirick, 2007), moral judgment (attitude), and remedies (behavior) in the direction set by the messages (Entman, 1993). Agendas, with a broad definition, linger among exposure, importance, and cognition, while frames are ways of organizing and interpreting cognitive, attitudinal, and behavioral components. Agenda-setting research addresses one small part, but the fundamental levels, of the complex interrelationship among media, public opinion, and public policy making (Kosicki, 1993). This distinction between their dependent variables can be captured if the hierarchy of response model of the audience is consulted in the debate.

Even their independent variables are different: media (message) agendas are measured with the amount of media coverage contributed to an issue or certain aspect of an issue, whereas media (message) frames focus more on different ways of constructing the meaning of the issue, namely, "clusters of messages" (Entman, 1993, p. 52) or "organized symbolic devices that will interact with individual agents' memory for meaning construction" (Pan & Kosicki, 1993, p. 53). Message frames usually contain components of moral or behavioral suggestions (Entman, 1993; Moy et al., 2016).

The difference in the variables of the message and audience agendas and frames can be illustrated with two studies on the media messages of, and public opinion about, the Afghanistan war, both using a combination of media content analysis and survey (see Table 5.1). Agendas are defined as sub-topics of the war, such as air traffic safety, and the Israel-Palestine Conflict (Craft & Wanta, 2004). Frames are defined as ways to describe the people involved in the 9/11 tragedy, the cause of the war. The war frame describes the dead in the tragedy as casualties and the perpetrators as enemies that "should be killed on the battlefield," while the crime frame portrays the dead as murder victims and the perpetrators as criminals that "should be tried in a court of law" (Edy & Meirick, 2007, p. 123).

Table 5.1 Difference between agendas and frames of the message and the audience

Study	*Agenda-setting (Craft & Wanta, 2004)*	*Framing effects (Edy & Meirick, 2007)*
IV (Message)	The number of stories on each subtopic.	War frame: the dead in 9/11 are casualties; perpetrators are enemies Crime frame: the dead in 9/11 are victims; perpetrators are criminals
DV (Audience)	Public rankings of each subtopic as a concern	War frame: perpetrators should be killed on the battlefield. Crime frame: perpetrators should be tried in court.

Frankly, differentiating second-level agenda-setting from framing effects has never been easy, because they are interconnected processes between the same parties (the communicator/message and the audience) and involve similar effects (Moy & Bosch, 2013; Scheufele & Tewksbury, 2007; Weaver, 2007; Weaver et al., 2004). So, while differentiation is needed for research precision (Scheufele & Tewksbury, 2007), a holistic approach to understanding them, or even more levels of communication effects, is also indispensable (Iyengar, 1991). Research has been conducted to develop a multilevel model of communication effects (Yao, 2008; Yao et al., 2020).

4 The Five-Level Model of Message Effects

Integrating the levels of communication effects that appear in the multi-level models (Johnson-Cartee, 2004; Lavidge & Steiner, 1961; Leckenby & Wedding, 1982; McGuire, 1989, 2013; McLeod & Reeves, 1980; McQuail, 2010; Potter, 2012) and the agenda-setting (Craft & Wanta, 2004; McCombs & Reynolds, 2009; McCombs & Shaw, 1993), priming (Roskos-Ewoldsen et al., 2002; Sheafer & Weimann, 2005), and framing (Edy & Meirick, 2007; Pan & Kosicki, 1993; Scheufele & Tewksbury, 2007) research, a parsimonious five-level model with attention, agenda (issue importance), knowledge (cognition), attitude, and behavior has been proposed and tested with environmental communications through mass media (Yao, 2008; Yao et al., 2020).

4.1 *The Five-Level Model*

In the five-level model of communication effects, when communication effects accumulate to pass a threshold at a lower level, the effects at the next higher level are activated (Leckenby & Wedding, 1982; McGuire, 1989, 2013). Meanwhile, message stimuli at one level have a stronger influence on the audience at the same level, and a higher-level process implies effects at levels below it. Effects at higher levels are usually harder to capture because they depend on the accumulation of effects at lower levels, which explains why agenda-setting is a robust finding among the communication effects (Yao et al., 2020). To simplify the discussion, this five-level model hypothesizes a one-way flow informational situation (no competitive content; Zaller, 1992) and focuses on individual audience members' short-term effects (McQuail, 2010), leaving the aggregated communication effects and long-term effects, such as McGuire's (1989) reinforcement and consolidation, for future research. Also left are moderators of communication effects, notably source credibility (McGuire, 1989), need for orientation (McCombs & Reynolds, 2009) or elaboration likelihood (Petty et al., 2002), and predisposition (Chong & Druckman, 2007), due to the scope of this chapter.

The model (see Fig. 5.2) is discussed level by level in the following section. Nine propositions are proposed to elaborate the model and tested with the NCSSLS data.

The Level of Attention

Attention, in the message, is the amount of information released; on the audience end, it is the amount of time and energy invested in exposure to the message. Attention is the lowest level without which both message stimuli and audience responses at higher levels cannot happen. Although some of the communication models we have discussed include a level of exposure (e.g., McGuire, 1989), more researchers believe that attention is a more meaningful initial level of a communication effect model because it represents attentive exposure (Chaffee & Schleuder, 1986; Yao et al., 2020). Therefore, Proposition 1:

P1 *The more attention the message gives to an issue, the more attention the audience will give to that issue.*

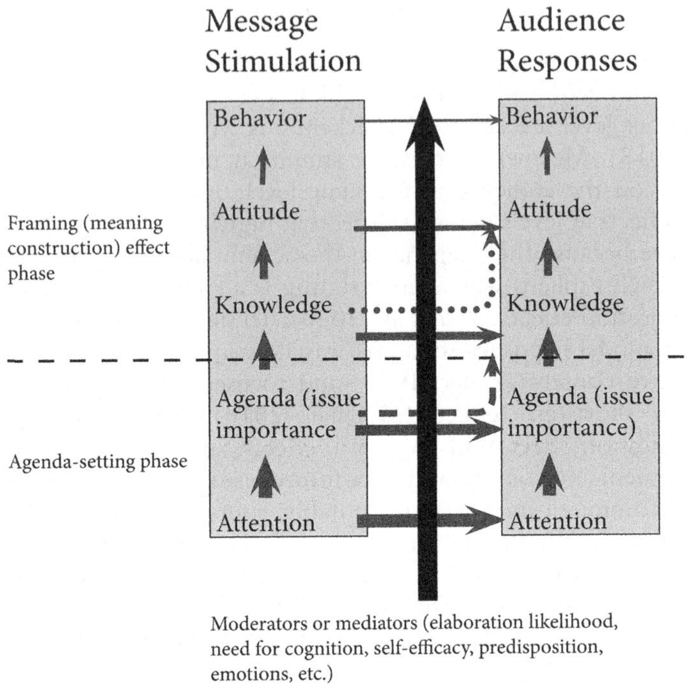

Fig. 5.2 The five-level model of message effects. (Notes: 1. As messages on an issue increase in levels—starting from attention to agenda (salience or importance), knowledge (cognitive recognition), attitude, and behavior—the persuasive effect on the audiences also increases at levels. The increases at higher levels of audience responses come from two sources: the message effects at the same level, and the effects accumulated from lower levels of audience responses. Hence, the higher the levels, the weaker the message effects. 2. The second level of message effects is the classical agenda-setting theory in media studies, which states that the media agenda leads to a positive change in the audience agenda. The dashed line represents the priming theory proposed by Iyengar and Kinder (1987), which extends the dependent variable of the classical agenda-setting theory to the criteria for evaluating politicians. The third level is the classical framing effects, although many framing effects studies adopt the dotted line path, measuring the message variable at the third level but the audience variable at a higher level. The studies of Lazarsfeld's team are at the fifth level, measuring the influence of messages' call for action on the audience's voting intentions. 3. For parsimony, the model does not address moderators in detail)

The Level of Agenda (Salience or Issue Importance)
Although message attention and message agenda are sometimes used interchangeably, the two concepts actually mean different things on both the communicator's end and the audience's end (Becker, 1982). Agenda, a great metaphor created by McCombs and Shaw (1972), refers to the issue importance (salience) in the message or the audience's mind. Communicators usually attach a degree of importance when they release a message. For instance, journalists always rank news according to its importance through particular arrangements (e.g., putting it on the front page) or by granting more attention (e.g., providing more space or greater length). In national media, for instance, the most conspicuous news stories usually "are nation and society—their persistence, cohesion, and the conflicts and divisions threatening their cohesion" (Gans, 1979, p. 19). Such salience or degree of importance is likely to increase when the communicator or the audience gives more attention to an issue. So, Proposition 2:

P2 *The more attention the communicator and the audience give to an issue, the more likely they think of that issue as important.*

These two propositions, however, cannot be tested in this book due to the limitation of the NCSSLS data used, which will be left for future research.

The message agenda (attached importance), rather than the message attention (amount of information released), is the stronger predictor of audience agenda. This communication effect is predicted in the classical first-level agenda-setting theory (McCombs & Reynolds, 2009; Moy et al., 2016; Moy & Bosch, 2013), which has been tested in one of the four perspectives described in the "Acapulco Typology" (McCombs & Reynolds, 2009), sometimes on environmental issues (e.g., Soroka, 2002; Trumbo, 1995). The fourth perspective, "cognitive portrait" (pp. 5–6), examines how the media agenda on a single item (object or attribute) impacts the subject's ranking of importance of that item, with an experimental design. Using that approach, Iyengar and Kinder (1987), for instance, find that stories at the top of the broadcasts are more influential in setting the public's agenda, no matter how long or short they are. Once the effects of those top stories are controlled, few effects of mass communication issue importance appear elsewhere (Yao, 2008). Hence Proposition 3, which is the first hypothesis of this book:

P3 *The more the message portrays an issue as important, the more likely the audience will think of that issue as important* (H1; classical agenda-setting).

The Level of Knowledge (Cognitive Information or Interpretation)

To address the criticism that agenda-setting research pays little attention to the textual information and cognitive content in the messages and their influences on the audience's mind (Edelstein, 1993; Iyengar & Kinder, 1987), more framing effect studies have been conducted (Reese, 2007) to examine the persuasion part of the communication content. Those framing effect studies start from the level of knowledge (Yao et al., 2020).

Perceiving some issues as important does not necessarily mean knowing those issues, as evidenced by some polls showing that most people think of global warming as important but cannot give correct answers in knowledge tests about global warming (Nisbet & Myers, 2007; Smith, 2002). The perception of importance, however, can lead to an increase in knowledge or trigger effects at even higher levels, for both the communicator and the audience, because when people perceive an issue as important, they will give more attention to it, gain more information about it from their communication channels, and are more likely to attend legislative activities to tackle that issue (Hutchings, 2005; Yao et al., 2020). So, Proposition 4:

P4 *The more the communicator and the audience think of an issue as important, the more knowledge they gain of that issue* (H2).

The fundamental elements of messages, even the highly persuasive ones, are cognitive information, organized in a certain way, which can shape the organization of cognitive elements on issues in the audience's mind (Yao et al., 2020). For instance, the audience that is exposed to messages portraying the security of water, energy, and food as a nexus is expected to think of these three resources as a nexus (Yao et al., 2018). Because information/knowledge is the fundamental constituent of the messages, scholars believe that Klapper (1960) would have found a strong media effect had he examined changes in the audience's knowledge rather than attitudes (Chaffee et al., 1970; Yao, 2008). This is also the prediction of the classical framing effect studies (Price et al., 1997; Shen, 2004; Shen & Edwards, 2005; Yao, 2007) and Proposition 5:

P5 *The more the message provides knowledge of an issue from a certain perspective, the more knowledge of that issue from that perspective the audience will have* (H3; classical frame-setting).

The Level of Attitude (Affection)
Messages often contain values, attitudinal components (Entman, 1993; Pan & Kosicki, 1993), or "the affective attributes" (Sheafer, 2007, p. 21; Yao, 2008), which are perhaps more apparent in opinion pieces but actually in all messages, maybe sometimes implicitly. Those affective or attitudinal factors are usually suggested in or nurtured by the message's cognitive frames, namely, the aspects chosen, or the ways the aspects are organized as an interpretation in the message (Hibbing & Theiss-Morse, 1998; Yao, 2007). Attitudes can also be directly developed from the accumulation of cognitive knowledge (Rogers, 2003). Political scientists' priming research (Iyengar & Kinder, 1987) mainly finds that the media agenda (without controlling media knowledge and attitude), through audience agenda and knowledge, influences the audience's standard of evaluation and hence its attitude (Yao et al., 2020), showing that "media emphasis on particular issues not only confers status (or increases salience), but also activates in people's memories previously acquired information about these issues," which "thus is then used in forming opinions about persons, groups, or institutions linked to these issues" (McCombs et al., 1991, p. 19). So, Proposition 6:

P6 *The more knowledge of an issue the communicator and the audience have, the more likely they both will form an attitude toward that issue consistent with their knowledge* (H4).

The affective or attitudinal elements based on or suggested in the cognitive elements in the messages, as intended by the communicators, can shape the audience's affection or attitude toward the issue or aspect in the message, as several empirical studies have found (Yao, 2008; Yao et al., 2020). A study of the relationship between news coverage and public opinion between 1969 and 1983 finds that "what appears on TV news," especially in the news commentaries and reported expert opinions (which usually carry heavy affective or attitudinal components), "accounts in large part for the relatively short-term (neither instantaneous nor glacial) changes in public opinion" (Page & Shapiro, 1992, p. 344). Another one also notices that newspaper editorials (also known for their saturation of explicitly affective or attitudinal components), compared with news stories, have a stronger positive correlation with public opinions six months later (Chang, 1992). The affective or attitudinal components in the messages can change their audience's affection or attitude toward the issue or aspect in those messages because those messages pull a particular consideration (containing both "cognitive elements" and "affects") to the more accessible position in their audience's minds, making this consideration easier to be expressed when the audience gives an opinion or expresses its affection or attitude on the issue or aspect (Zaller, 1992, p. 59). This relationship between the messages and the audience at the affective or attitudinal level, also termed "affective agenda-setting" (e.g., Coleman & Wu, 2022) or "affective framing" (e.g., Greenaway & Fielding, 2020) in some other studies, is Proposition 7:

P7 *The stronger the message attitude toward an issue, the stronger the audience attitude toward that issue, in the same direction* (H5).

The Level of Behavior

Messages, in addition to providing cognitive information in a certain interpretive way and portraying issues or aspects with certain moral, affective, or attitudinal angles, often also contain behavioral suggestions, such as remedy recommendations (Entman, 1993), violence description (Roskos-Ewoldsen et al., 2002), or even directly call to action (Yao, 2021). The audience members, once forming their own frames based on media frames, are also likely to form their attributions of responsibility, attitudes, and therefrom behaviors (Fishbein & Ajzen, 2010; Scheufele, 1999). So, Proposition 8:

P8 *The stronger the communicator and the audience have an attitude toward an issue, the more likely they both will call for actions or act toward that issue in a way consistent with their attitudes* (H6).

Behavioral suggestions or calls in the messages, verbal or visual, can lead to similar behaviors in the audience (Roskos-Ewoldsen et al., 2002; Yao, 2021). The most intensive evidence of communication effects at the behavioral level exists in media violence studies, which consistently find that viewing media violence increases subsequent aggressive behaviors (Bandura et al., 1963; Berkowitz & Rawlings, 1963). A meta-analysis of 217 studies of media violence from 1957 to 1990 concludes that the amount of violent television viewing is positively associated with the degree of subsequent aggressive behaviors (Paik & Comstock, 1994). A body of media sexuality literature also suggests that consuming media content with behavioral components elicits copy acts (Harris & Scott, 2002). A study of the COVID-19 pandemic also finds that viewers of *Hannity*, a popular news show on Fox News that initially framed the pandemic as an insignificant threat, adopted preventative behaviors later and were more likely infected or killed by the disease than viewers of *Tucker Carlson Tonight*, another popular news show on the same channel (Bursztyn et al., 2020). Lazarsfeld's two studies were at the behavioral level too, which is why the communication effects were harder to capture than they expected. Communication effects at this behavioral level are also the focus of psychological priming research (Roskos-Ewoldsen et al., 2002; Yao et al., 2020) and social learning theory (Bandura, 1977), from different angles. So, Proposition 9:

P9 *Message advocacy of a certain behavior increases the probability that the audience conducts the same behavior* (H7; psychological priming or social learning theory).

In the following section, these propositions are tested with the second round of the National College Student Science Literacy Survey (NCSSLS) data on the topic of the water-energy-food (WEF) nexus.

4.2 Testing the Five-Level Model of Message Effects on the WEF Nexus

To test the levels of message effects hypothesized above, five groups (control, agenda version, knowledge version, attitudinal version, and the fourth behavioral version; all message versions one-sided, closed-ended, and gain-framed to eliminate confounding) were coded from the second round of the NCSSLS data, which included 149 responses from 102 universities across the U.S. The versions of the WEF-nexus message were differentiated by consisting of only a conspicuous appearance (as the agenda version with only a large, long title), a knowledge component, an attitudinal component, or a behavioral component, with a higher level of version always containing the components at the lower levels of the five-level model of message effects (see Chap. 4 and Appendix; Yao et al., 2020).

The Statistical Results of the Testing
The subjects were aged 18–63. Sophomores accounted for 12.9%, juniors 21.4%, seniors 37.9%, master's students 13.6%, doctoral students 9.3%, all-but-dissertation doctoral candidates 2.9%, and other 2.1%. Among the subjects, 32 were exposed to no message, 26 to the agenda version, 27 to the knowledge version, 31 to the attitudinal version, and 33 to the behavioral (call for action) version.

Testing the Levels of Message Effects on the WEF Nexus
To test H1, the classic agenda-setting theory in the case of the WEF nexus, a hierarchical multiple regression model of subjects' WEF-nexus agenda was built, with predictors being age, education, keeping up with science news, subjective WEF-nexus knowledge 1, answer quality, and a dummy variable of whether being exposed to a message version. Subjects' WEF-nexus agenda, that is, the importance that the subjects granted to the discussion whether the water, energy, and food security should be managed as a nexus, showed no difference between those exposed to the agenda, knowledge, attitudinal, or behavioral version and those in the control group. The coefficient of the group being exposed to a message version ($\beta = -0.14$) was surprisingly negative and the p-value was smaller than 0.10 (a cutoff p-value used in some social science work, such as Iyengar & Kinder, 1987), not very far away from the traditional cutoff value of 0.05 (see Table 5.2), indicating the audience's reactance to persuasion. So, not only was the agenda-setting hypothesis (H1) not confirmed with these data, but it was also almost rejected. Noticeably, however, keeping up with science news showed a substantial positive association with subjects' agenda on the WEF nexus ($\beta = 0.35$, $p < 0.001$), even with other variables along in the model, suggesting a robust agenda-setting effect over a longer time period.

Three more hierarchical multiple regression models with the same set of five control variables (age, education, keeping up with science news, subjective WEF knowledge 1, and answer quality) were conducted to test the hypotheses H3, H5, and H7. Three dummy variables of whether being exposed to a version containing the informational component, attitudinal component, and the behavioral component, respectively, were set as the major predictors in those hierarchical regression models. Only the hypothesis at the knowledge level (H3) was confirmed at the conventional level of $\alpha = 0.05$ (see Table 5.2). The objective WEF-nexus knowledge of those subjects who were exposed to versions with knowledge components

Table 5.2 Hierarchical regression models of WEF-nexus agenda, objective knowledge (and frame), attitude, and behavioral intention on the message stimulus at levels

IV		*Agenda*	*Objective knowledge*	Frame	*Attitude*	*Behavior*
Block 1	*Age*	0.11	0.08	0.12	0.15	**0.23****
	Education	−0.12	**0.18****	0.006	−0.15	−0.03
Block 2	*Keeping up with science news*	**0.35******	**0.21****	0.07	**0.30*****	**0.20****
	Subjective Knowledge 1	0.02	0.10	**0.20****	0.005	0.14
	Answer quality	0.14*	**0.25*****	0.13	0.15*	0.12
Block 3	*Exposed to the agenda, Knowledge, attitudinal, or behavioral message (= 1)*	−0.14*	—	—	—	—
	Exposed to the knowledge, attitudinal, or behavioral message (= 1)	—	**0.18****	−0.15*	—	—
	Exposed to the attitudinal or behavioral message (= 1)	—	—	—	0.04	—
	Exposed to the behavioral message (= 1)	—	—	—	—	0.06
Model info		$R^2 = 0.16$, Adj. $R^2 = 0.12$, all *VIFs* < 1.27.	$R^2 = 0.28$, Adj. $R^2 = 0.24$, all *VIFs* < 1.26.	$R^2 = 0.10$, Adj. $R^2 = 0.06$, all *VIFs* < 1.30.	$R^2 = 0.14$, Adj. $R^2 = 0.10$, all *VIFs* < 1.30.	$R^2 = 0.16$, Adj. $R^2 = 0.12$, all *VIFs* < 1.30.

Listed are the final models of the hierarchical multiple regressions

Entries are betas, with * denoting $p < 0.1$, ** denoting $p < 0.05$, *** denoting $p < 0.01$, and **** denoting $p < 0.001$. — means the variable was not entered into the model (so each model has six IVs). Coefficients statistically significant at the traditional standard ($p < 0.05$) are bolded

Missing values replaced with means

(the knowledge, attitudinal, and behavioral versions) was higher than those who were not exposed to a version with the knowledge component (in the control group or the group exposed to the agenda version. $\beta = 0.18$, $p < 0.05$). This is consistent with the finding of the extant studies that the communication effects are easier to identify at the knowledge level (Chaffee et al., 1970). Meanwhile, more education ($\beta = 0.18$, $p < 0.05$), more keeping up with science news ($\beta = 0.21$, $p < 0.05$), and higher survey-taking quality ($\beta = 0.25$, $p < 0.01$) were also associated with a higher score on the objective WEF-nexus knowledge test. The hypothesis with the WEF-nexus frame (also H3), however, was almost refuted. Being exposed to a version with the knowledge component seemed to negatively contribute to subjects' recognition of water, energy, and food as a nexus ($\beta = -0.15$, $p < 0.1$). Subjects' subjective WEF-nexus knowledge 1, however, as a control variable was positively associated with their frame of the WEF nexus ($\beta = 0.20$, $p < 0.05$).

Not confirmed was the hypothesis predicting that the groups of subjects who were exposed to a WEF message with the attitudinal components tended to have a higher supporting attitude toward the WEF-nexus approach than the groups of subjects who were exposed to no message or a version without attitudinal components (H5). Keeping up with science news ($\beta = 0.30$, $p < 0.01$) was found positively associated with subjects' supportive attitude toward the WEF nexus (suggesting a long-term framing effect), and possibly also the quality of subjects' survey responses ($\beta = 0.15$, $p < 0.1$).

Also not confirmed was the hypothesis predicting that the group exposed to the WEF-nexus message with behavioral calls tended to show higher intention to do something to support the WEF-nexus approach, compared to the groups exposed to no message or a version without the behavioral component (H7). The hierarchical multiple regression model of WEF-nexus behavioral intention showed only age ($\beta = 0.23$, $p < 0.05$) and, again, keeping up with science news ($\beta = 0.20$, $p < 0.05$) as statistically significant positive predictors of subjects' supportive behavioral intention.

Five MANCOVA analyses were conducted, using WEF-nexus agenda, frame, objective knowledge, attitude, and behavioral intention as dependent variables, the group variable (four message groups and the control group) as the fixed factor, and the five control variables (age, education, keeping up with science news, subjective WEF knowledge 1, and answer quality) as covariates. They showed some differences in the five response variables across the five groups in the subjects' WEF-nexus frame and objective knowledge. The post hoc LSD comparison showed that subjects who were exposed to the behavioral version ($M_{difference} = -0.72$, $p = 0.015$) and possibly the attitudinal version ($M_{difference} = -0.50$, $p = 0.09$) were less likely to recognize water, energy, and food as a nexus than subjects in the control group, revealing psychological reactance in the audience once again. The group exposed to the attitudinal version, however, had a higher score in objective WEF nexus knowledge than the control group ($M_{difference} = 0.66$, $p = 0.035$), and possibly so did the group exposed to the agenda version ($M_{difference} = 0.59$, $p = 0.07$). Subjects who were exposed to the knowledge version also possibly performed better in the knowledge test than the control group ($M_{difference} = 0.62$, $p = 0.06$). Those MANCOVA analyses also confirm that knowledge is a communication effect that is easier to be identified (Chaffee et al., 1970).

Testing the Hierarchy of Audience Responses on the WEF Nexus
To test whether positive relationships exist among subjects' agenda, knowledge (objective knowledge; frame also measured), attitude, and behavioral intention on the WEF nexus (H2, H4, H6), zero-order and partial correlations (controlling age, education, keeping up with science news, subjective WEF-nexus knowledge 1, and answer quality) were conducted.

As shown in Table 5.3, the hypothesis predicting a linkage between the audience's WEF-nexus agenda and its objective WEF-nexus knowledge (H2) was confirmed by the zero-order correlation analysis but not the partial correlation analysis. The predicted linkage between the objective WEF-nexus knowledge and WEF-nexus attitude (H4) was not confirmed by either correlation analysis with data from the second round (it was, however, supported by both the zero-order analysis and the partial analysis in the first round of the data. Please refer to Chap. 4). The predicted linkage between the audience's attitude and behavioral intention toward the WEF nexus (H6) was confirmed by both the zero-order and the partial correlation analyses. Substituting the subjects' objective knowledge of the WEF nexus with their frame on the WEF nexus, then all three hypotheses were confirmed by both correlation analyses, showing that the WEF-nexus frame could be a better concept than objective WEF-nexus knowledge in the cognitive-affective-conative hierarchy (Yao et al., 2020). As stated in H2, subjects who thought that it was important to discuss whether the water, energy, and food resources should be managed as a nexus (WEF-nexus agenda) tended to score high in the recall test of

Table 5.3 Zero-order and partial correlations among subjects' agenda, objective knowledge (and frame), attitude, and behavioral intention on the WEF nexus

		Agenda	*Objective knowledge*	*Frame*	*Attitude*
Objective knowledge	Zero-order	0.29 (<0.001, 143)			
	Partial	—			
Frame	Zero-order	0.27 (0.001, 141)	—		
	Partial	0.20 (0.025, 121)	—		
Attitude	Zero-order	0.66 (<0.001, 139)	—	0.51 (<0.001, 139)	
	Partial	0.60 (<0.001, 121)	—	0.53 (<0.001, 121)	
Behavioral intention	Zero-order	0.51 (<0.001, 141)	—	0.32 (<0.001, 141)	0.69 (<0.001, 139)
	Partial	0.46 (<0.001, 121)	—	0.22 (0.013, 121)	0.66 (<0.001, 121)

Missing data were excluded pairwise

For the partial correlations, keeping up with science news, subjective WEF-nexus knowledge 1, answer quality, age, and education were in control

Entries are, in order, *r* (*p*, *n*) for the zero-order correlations and *r* (*p*, *df*) for the partial correlations. Correlations to self and repeat correlations not reported; insignificant correlations marked with —

knowledge points from the informational section of the WEF-nexus message (objective WEF-nexus knowledge; zero-order: $r = 0.29$, $p < 0.001$, $n = 143$) and also tended to recognize that water, energy, and food are indeed a nexus (WEF-nexus frame; zero-order: $r = 0.27$, $p = 0.001$, $n = 141$; partial: $r = 0.20$, $p = 0.025$, $df = 121$). Although subjects who scored higher in the knowledge test about the WEF-nexus message showed no tendency to think of the WEF-nexus concept as important, useful, or urgent to adopt to tackle serious issues, those who recognized water, energy, and food security as a nexus were more likely to do so (zero-order: $r = 0.51$, $p < 0.001$, $n = 139$; partial: $r = 0.53$, $p < 0.001$, $df = 121$) as predicted in H4 and by the five-level model of message effects (Yao et al., 2020). The predicted positive relationship between the audience's WEF-nexus attitude and WEF-nexus behavioral intention (H6) was also found substantial (zero-order: $r = 0.69$, $p < 0.001$, $n = 139$; partial: $r = 0.66$, $p < 0.001$, $df = 121$).

Interpreting the Five Levels of Media Effects on the WEF Nexus

The results of those analyses overall support the idea that messages on new environmental topics such as the WEF nexus are processed in the audience's mind at different levels and that the levels are connected in a hierarchical manner (McGuire, 1989, 2013; Rogers, 2003; Yao et al., 2020). The audience's WEF-nexus agenda, that is, people's thinking of a topic like the WEF nexus as important to discuss, is positively associated with the audience's recall of information about that topic (e.g., objective WEF-nexus knowledge), recognition of the way to interpret the topic that is presented in the messages (e.g., WEF-nexus frame), as well as the audience's attitude, and behavioral intentions toward the issue (e.g., WEF-nexus attitude and behavioral intention). The audience's objective WEF-nexus knowledge as recall of information, although not being related to other concepts in the hierarchy of audience responses (Lavidge & Steiner, 1961; Leckenby & Wedding, 1982; McGuire, 1989; Ray, 1973; Rogers, 2003) as shown by the NCSSLS data, still quite possibly positively correlates to WEF-nexus attitude and behavioral intentions (as shown in the analysis of the data collected in the first round, which has better representativeness), just as do the two subjective measurements of WEF-nexus knowledge. Furthermore, in the second round of the NCSSLS data, the audience's WEF-nexus frame, a more specific measurement of the audience knowledge, perfectly substitutes the place of the audience's objective WEF-nexus knowledge and positively correlates to the audience's WEF-nexus agenda at the lower part of the hierarchy of audience responses and the audience's WEF-nexus attitude at the upper part of the hierarchy. Data of the second round also clearly show that the audience's WEF-nexus attitude and audience's WEF-nexus behavioral intentions are positively associated. So, as described in the classical hierarchy of response model (Lavidge & Steiner, 1961; Leckenby & Wedding, 1982; McGuire, 1989; Ray, 1973; Rogers, 2003;Yao, 2008 ; Yao et al., 2020), audience responses are stimulated by messages at levels. The accumulation of responses at a lower level will activate the next higher level when the volume of the responses at the lower level passes a threshold (Yao et al., 2020).

Although only the knowledge level of message effect, among the five levels of message effects hypothesized based on the five-level model of message effects (including the level of frame as a special form of the level of knowledge; Yao et al., 2020), is confirmed with the second round of the

NCSSLS data, the five-level model of message effects, integrated based on a synthetic review of existing studies of message effects from multiple areas of studies, should not be abandoned but needs further testing and modification, as the pattern of the data is still promising. Meanwhile, the conspicuous positive contribution of "keeping up with science news" as a variable to the audience's WEF-nexus agenda, objective WEF-nexus knowledge, WEF-nexus attitude, and WEF-nexus behavioral intention, which has been identified in the tests of hypotheses in this chapter, supports the idea that communication effects happen as hypothesized in this chapter in the long term. Evidence from one study would not approve a theoretical model, nor could the model be disapproved without clear evidence of rejection. Further tests are needed based on the tests in this chapter to cast more light on our understanding of message stimuli and effects at different levels.

Also worthy of attention are agenda and frame (for both the communicators and the audience), two concepts contributed by mass communication scholars (Cacciatore et al., 2016; McCombs & Reynolds, 2009; McCombs & Shaw, 1993; Yao, 2007) and integrated into the five-level of message effects (Yao, 2008; Yao et al., 2020). The NCSSLS data show that both agenda and frame need further explication. The audience's WEF-nexus frame, measured as recognizing the way that the messages interpret the WEF nexus, is found a better predictor of both WEF-nexus attitude and WEF-nexus behavioral intention, compared with objective WEF-nexus knowledge. It is also positively related to the audience's WEF-nexus agenda, the thought that the WEF-nexus approach is important to discuss no matter whether an agreement with the message interpretation exists or not. The audience's WEF-nexus agenda, in fact, also positively relates to the subjects' WEF-nexus attitude and WEF-nexus behavioral intention, perhaps more strongly than the subjects' WEF-nexus frame (a Fisher r-to-z test showed a one-tailed $p = 0.034$ for the difference between the agenda-attitude and frame-attitude correlation-coefficients and a one-tailed $p = 0.026$ for the difference between the agenda-behavioral intention and the frame-behavioral intention coefficients, with the former coefficients in those two pairs of correlations being larger). So, the WEF-nexus message's agenda and the subjects' WEF-nexus agenda, although both termed as "agenda" by mass communication scholars (e.g.,McCombs & Reynolds, 2009 ; McCombs & Shaw, 1993), seem to contain different components. On the message side, measured as either the number of appearances or the length of the message (McCombs & Reynolds, 2009;

McCombs & Shaw, 1993), the message agenda assumes no attitudinal components. Mere message appearance with little knowledge, attitudinal, or behavioral components might decrease the messages' persuasive effects at the knowledge, attitudinal, and behavioral levels, as suggested by the sample means in the MANCOVA model, causing psychological reactance to persuasion in the audience. On the audience end, however, recognizing an issue as important to discuss implies high attitudinal and behavioral suggesting components, at least for the new topic of WEF nexus. If the topic of discussion is a complicated and controversial issue, such as climate change, maybe the audience can differentiate the importance of discussing the topic and the importance of adopting the idea sponsored in the messages, which may then make issue importance a more appropriate measurement of audience (public) agenda (notice that classical agenda-setting studies are mostly about highly debated issues emerged in presidential or national elections). So, with a complicated topic, the positive relationship between the audience's agenda and attitude, as well as the audience's behavioral intention, may be lowered or even disappear. In this study with the topic of WEF nexus, the high correlation between the audience's agenda and attitude, even stronger than the linkage between the audience's frame and attitude, suggests that a more accurate explication of the widely used concepts such as agenda, frame, and attitude need further scrutinization and explication.

Finally, the WEF nexus, an approach to managing water, energy, and food resources in a way widely considered by scientists to be more effective, is in urgent need of research and discussion (Biggs et al., 2015; International Institute for Sustainable Development, 2013; United Nations, 2014). Although still consisting of students, the sample for this study was randomly drawn from a variety of more than 100 colleges scattered in the U.S., providing greater representativeness and external validity. The survey-experimental design also provides better generalizability with its real-life setting. The trade-off, however, is that more errors are contained in the data than those collected in a strictly controlled lab, which may have obscured or even distorted some of the hypothesized effects. Of course, the small magnitude of the single-message stimulus embedded in the survey experiment could also have been responsible for not confirming the message effects at several levels, even including the classic and robust theory of agenda-setting (McCombs, 2005), whose opposite is almost supported, the hypothesis of frame-setting (Edy & Meirick, 2007; Scheufele, 1999), and the behavioral influences

(Roskos-Ewoldsen et al., 2002). Clear evidence of long-term message effects, however, has been found at the knowledge and other levels. The model's advice that strategic environmental communicators should provide a plethora of cognitive information about the WEF nexus to the public before adding attitudinal and call-for-action components to their promotional messages should be valuable.

References

Bandura, A. (1977). *Social learning theory.* Prentice Hall.

Bandura, A., Ross, D., & Ross, S. A. (1963). Imitation of film-mediated aggressive models. *Journal of Abnormal and Social Psychology, 66*(1), 3–11. https://doi.org/10.1037/h0048687

Barry, T. E. (2002). In defence of the hierarchy of effects: A rejoinder to Weilbacher. *Journal of Advertising Research, 42*(3), 44–47. https://doi.org/10.2501/JAR-42-3-44-47

Becker, L. B. (1982). The mass media and citizen assessment of issue importance: A reflection on agenda-setting research. In D. C. Whitney & E. Wartella (Eds.), *Mass Communication Review Yearbook 3* (pp. 521–536). Sage.

Belch, G. E., & Belch, M. A. (2014). *Advertising and promotion: An integrated marketing communications perspective* (10th ed.). McGraw-Hill/Irwin.

Berkowitz, L., & Rawlings, E. (1963). Effects of film violence on inhibitions against subsequent aggression. *Journal of Abnormal and Social Psychology, 66*(5), 405–412. https://doi.org/10.1037/h0046639

Besley, J., & Dudo, A. (2022). *Strategic science communication: A guide to setting the right objectives for more effective public engagement.* Johns Hopkins University Press.

Bettinghaus, E. P. (1986). Health promotion and the knowledge-attitude-behavior continuum. *Preventive Medicine, 15*(5), 475–491. https://doi.org/10.1016/0091-7435(86)90025-3

Biggs, E., Bruce, E., Boruff, B., Duncan, J. M. A., Horsley, J., Pauli, N., McNeill, K., Neef, A., Van Ogtrop, F., Curnow, J., Haworth, B., Duce, S., & Imanari, Y. (2015). Sustainable development and the water-energy-food nexus: A perspective on livelihoods. *Environmental Science & Policy, 54*, 389–397. https://doi.org/10.1016/j.envsci.2015.08.002

Borah, P. (2011). Conceptual issues in framing theory: A systematic examination of a decade's literature. *Journal of Communication, 61*(2), 246–263. https://doi.org/10.1111/j.1460-2466.2011.01539.x

Bursztyn, L., Rao, A., Roth, C., & Yanagizawa-Drott, D. (2020). Misinformation during a pandemic. *Working Paper, 2020–44.* Becker Friedman Institute for Economics at the University of Chicago. https://bfi.uchicago.edu/wp-content/uploads/BFI_WP_202044.pdf

Cacciatore, M., Scheufele, D., & Iyengar, S. (2016). The end of framing as we know it... and the future of media effects. *Mass Communication and Society, 19*(1), 7–23. https://doi.org/10.1080/15205436.2015.1068811

Chaffee, S., & Lieberman, D. (2001). The challenge of writing the literature review: Synthesizing research for theory and practice. In A. Alexander & W. J. Potter (Eds.), *How to publish your communications research: An insider's guide* (pp. 23–46). Sage Publications.

Chaffee, S., & Schleuder, J. (1986). Measurement and effects of attention to media news. *Human Communication Research, 13*(1), 76–107. https://doi.org/10.1111/j.1468-2958.1986.tb00096.x

Chaffee, S., Ward, S., & Tipton, L. P. (1970). Mass communication and political socialization. *Journalism Quarterly, 47*(4), 647–666. https://doi.org/10.1177/107769907004700401

Chang, T. (1992). *The press and China policy: The illusion of Sino-American relations, 1950–1984.* Ablex Publishing Corporation.

Chong, D., & Druckman, J. N. (2007). Framing theory. *Annual Review of Political Science, 10*, 103–126. https://doi.org/10.1146/annurev.polisci.10.072805.103054

Coleman, R., & Wu, H. D. (2022). Individual differences in affective agenda-setting: A cross-sectional analysis of three U.S. presidential elections. *Journalism, 23*(5), 992–1009. https://doi.org/10.1177/1464884921990242

Craft, S., & Wanta, W. (2004). U.S. public concerns in the aftermath of 9-11: A test of second level agenda-setting. *International Journal of Public Opinion Research, 16*(4), 456–463. https://doi.org/10.1093/ijpor/edh039

Danielian, L., & Reese, S. (1989). A closer look at intermedia influence on agenda-setting: The cocaine issue of 1986. In P. Shoemaker (Ed.), *Communication campaigns about drugs: Government, media and the public* (pp. 47–66). Lawrence Erlbaum Associates.

Druckman, J. N. (2001). On the limits of framing effects: Who can frame? *The Journal of Politics, 63*(4), 1041–1066. https://doi.org/10.1111/0022-3816.00100

Druckman, J. N. (2004). Political preference formation: Competition, deliberation, and the (ir)relevance of framing effects. *American Political Science Review, 98*(4), 671–686. https://doi.org/10.1017/S0003055404041413

Druckman, J. N., & Nelson, K. R. (2003). Framing and deliberation: How citizens' conversion limit elite influence. *American Journal of Political Science, 47*(4), 729–745. https://doi.org/10.1111/1540-5907.00051

Edelstein, A. S. (1993). Thinking about the criterion variable in agenda-setting research. *Journal of Communication, 43*(2), 85–99. https://doi.org/10.1111/j.1460-2466.1993.tb01264.x

Edy, J. A., & Meirick, P. C. (2007). Wanted, dead or alive: Media frames, frame adoption, and support for the war in Afghanistan. *Journal of Communication, 57*(1), 119–141. https://doi.org/10.1111/j.1460-2466.2006.00332.x

Entman, R. M. (1993). Framing: Toward clarification of a fractured paradigm. *Journal of Communication, 43*(4), 51–58. https://doi.org/10.1111/j.1460-2466.1993.tb01304.x
Fishbein, M., & Ajzen, I. (2010). *Predicting and changing behavior: The reasoned action approach.* Psychology Press (Taylor & Francis).
Fishman, M. (1980). *Manufacturing the news.* University of Texas Press.
French, J., & Russell-Bennett, R. (2015). A hierarchical model of social marketing. *Journal of Social Marketing, 5*(2), 139–159. https://doi.org/10.1108/JSOCM-06-2014-0042
Gamson, W. A., & Modigliani, A. (1989). Media discourse and public opinion on nuclear power. *American Journal of Sociology, 95*(1), 1–37. https://doi.org/10.1086/229213
Gans, H. J. (1979). *Deciding what's news: A study of CBS Evening News, NBC Nightly News, Newsweek and Time.* Pantheon Books.
Gans, H. J. (1993). Reopening the black box: Toward a limited effects theory. *Journal of Communication, 43*(4), 29–35. https://doi.org/10.1111/j.1460-2466.1993.tb01301.x
Greenaway, T., & Fielding, K. S. (2020). Positive affective framing of information reduces risk perception and increases acceptance of recycled water. *Environmental Communication, 14*(3), 391–402. https://doi.org/10.1080/17524032.2019.1680408
Harris, R. J., & Scott, C. L. (2002). Effects of sex in the media. In J. Bryant & D. Zillmann (Eds.), *Media effects: Advances in theory and research* (2nd ed., pp. 307–331). Lawrence Erlbaum.
Hart, P. S., & Nisbet, E. C. (2012). Boomerang effects in science communication: How motivated reasoning and identity cues amplify opinion polarization about climate mitigation policies. *Communication Research, 39*(6), 1–23. https://doi.org/10.1177/0093650211416646
Hassan, S., Nadzim, S., & Shiratuddin, N. (2015). Strategic use of social media for small business based on the AIDA model. *Procedia: Social and Behavioral Science, 172*, 262–269. https://doi.org/10.1016/j.sbspro.2015.01.363
Hibbing, J. R., & Theiss-Morse, E. (1998). The media's role in public negativity toward Congress: Distinguishing emotional reactions and cognitive evaluations. *American Journal of Political Science, 42*(2), 475–498. https://doi.org/10.2307/2991767
Hutchings, V. L. (2005). *Public opinion and democratic accountability: How citizens learn about politics.* Princeton University Press.
International Institute for Sustainable Development. (2013). The water–energy–food security nexus: Towards a practical planning and decision-support framework for landscape investment and risk management. *IISD Website.* Retrieved March 4, 2025, from https://www.iisd.org/system/files/publications/wef_nexus_2013.pdf

Iyengar, S. (1991). *Is anyone responsible: How television frames political issues*. The University of Chicago Press.

Iyengar, S., & Kinder, D. R. (1987). *News that matters: Television and American opinion*. The University of Chicago Press.

Johnson-Cartee, K. S. (2004). *News narratives and news framing: Constructing political reality*. Rowman & Littlefield Publishers.

Kahneman, D., & Tversky, A. (1984). Choice, values, and frames. *American Psychologist, 39*(4), 341–350. https://doi.org/10.1037/0003-066X.39.4.341

Katz, E. (2001). Lazarsfeld's map of media effects. *International Journal of Public Opinion Research, 13*(3), 270–279. https://doi.org/10.1093/ijpor/13.3.270

Keller, K. (2012). *Strategic brand management: Building, measuring, and managing brand equity* (4th ed.). Pearson/Prentice Hall.

Klapper, J. (1960). *The effects of mass communication*. The Free Press.

Kollmuss, A., & Agyeman, J. (2002). Mind the gap: Why do people act environmentally and what are the barriers to pro-environmental behavior? *Environmental Educational Research, 8*(3), 239–260. https://doi.org/10.1080/13504620220145401

Kosicki, G. M. (1993). Problems and opportunities in agenda-setting research. *Journal of Communication, 43*(2), 100–127. https://doi.org/10.1111/j.1460-2466.1993.tb01265.x

Kotler, P., & Keller, K. (2006). *Marketing management* (12th ed.). Pearson/Prentice Hall.

Lavidge, R. J., & Steiner, G. A. (1961). A model for predictive measurements of advertising effectiveness. *Journal of Marketing, 25*(6), 59–62. https://doi.org/10.1177/002224296102500611

Lazarsfeld, P., Berelson, B., & Gaudet, H. (1944). *The people's choice: How the voter makes up his mind in a presidential campaign*. Duell, Sloan & Pearce.

Lazarsfeld, P., & Merton, R. K. (1948). Mass communication, popular taste, and organized social action. In L. Bryson (Ed.), *The communication of ideas* (pp. 95–118). Harper and Brothers.

Leckenby, J., & Wedding, N. (1982). *Advertising management: Criteria, analysis, and decision making*. Grid Publishing.

Lee, S., Song, H., Lee, C.-K., & Petrick, J. F. (2018). An integrated model of pop culture fans' travel decision-making processes. *Journal of Travel Research, 57*(5), 687–701. https://doi.org/10.1177/0047287517708619

Lippmann, W. (1922/1997). *Public opinion*. Free Press.

McCombs, M. (2005). A look at agenda-setting: Past, present, and future. *Journalism Studies, 6*(4), 543–557. https://doi.org/10.1080/14616700500250438

McCombs, M., Einsiedel, E., & Weaver, D. (1991). *Contemporary public opinion: Issues and the news*. Routledge.

McCombs, M., & Reynolds, A. (2009). News influence on our pictures of the world. In J. Bryant & D. Zillmann (Eds.), *Media effects: Advances in theory and research* (3rd ed., pp. 1–16). Routledge.

McCombs, M., & Shaw, D. (1972). The agenda-setting function of mass media. *Public Opinion Quarterly, 36*(2), 176–187. https://doi.org/10.1086/267990

McCombs, M., & Shaw, D. (1993). The evolution of agenda-setting research: Twenty-five years in the marketplace of ideas. *Journal of Communication, 43*(2), 58–67. https://doi.org/10.1111/j.1460-2466.1993.tb01262.x

McGuire, W. J. (1989). Theoretical foundations of campaigns. In R. E. Rice & C. K. Atkin (Eds.), *Public communication campaigns* (2nd ed., pp. 43–65). Sage.

McGuire, W. J. (2013). McGuire's classic input-output framework for constructing persuasive messages. In R. E. Rice & C. K. Atkin (Eds.), *Public communication campaigns* (4th ed., pp. 133–146). Sage.

McKenzie-Mohr, D., Lee, N. R., Schultz, W. P., & Kotler, P. (2012). *Social marketing to protect the environment: What works.* Sage Publications.

McLeod, J., Pan, Z., & Rucinski, D. (1995). Levels of analysis in public opinion research. In T. L. Glasser & C. T. Salmon (Eds.), *Public opinion and the communication of consent* (pp. 55–85). The Guilford Press.

McLeod, J., & Reeves, B. (1980). On the nature of mass media effects. In S. B. Withey & R. P. Abeles (Eds.), *Television and social behavior: Beyond violence and children: A report of the committee on television and social behavior social science research council* (pp. 17–54). Lawrence Erlbaum.

McQuail, D. (2010). *McQuail's mass communication theory* (6th ed.). Sage Publications.

Miller, J. M., & Krosnick, J. A. (2000). News media impact on the ingredients of presidential evaluations: Politically knowledgeable citizens are guided by a trusted source. *American Journal of Political Science, 44*(2), 301–315. https://doi.org/10.2307/2669312

Miller, S. (2001). Public understanding of science at the crossroads. *Public Understanding of Science, 10*(1), 115–120. https://journals.sagepub.com/doi/abs/10.3109/a036859

Moy, P., & Bosch, B. (2013). Theories of public opinion. In P. Cobley & P. J. Schultz (Eds.), *Handbook of communication science: Theories and models of communication* (Vol. 1, pp. 289–308). De Gruyter Mouton.

Moy, P., Tewksbury, D., & Rinke, E. M. (2016). Agenda-setting, priming, and framing. In K. B. Jensen & R. T. Craig (Eds.), *The international encyclopedia of communication theory and philosophy* (Vol. 1, pp. 52–64). Wiley. https://doi.org/10.1002/9781118766804.wbiect266

Mutz, D. C. (1998). *Impersonal influence: How perceptions of mass collections affect political attitudes.* Cambridge University Press.

Nisbet, M. C., & Myers, T. (2007). The polls – Twenty years of public opinion about global warming. *The Public Opinion Quarterly, 71*(3), 1–27. https://doi.org/10.1093/poq/nfm031

O'Keefe, D. J., & Jensen, J. D. (2006). The advantages of compliance or the disadvantages of noncompliance? A meta-analytic review of the relative persuasive effectiveness of gain-framed and loss-framed messages. In C. S. Beck (Ed.), *Communication yearbook* (pp. 1–43). Lawrence Erlbaum.

Page, B. I., & Shapiro, R. Y. (1992). *The rational public: Fifty years of trends in Americans' policy preferences.* The University of Chicago Press.

Paik, H., & Comstock, G. (1994). The effects of television on antisocial behavior: A meta-analysis. *Communication Research, 21*(4), 516–546. https://doi.org/10.1177/009365094021004004

Pan, Z., & Kosicki, G. M. (1993). Framing analysis: An approach to news discourse. *Political Communication, 10*(1), 55–76. https://doi.org/10.1080/10584609.1993.9962963

Pember, D., & Calvert, C. (2008). *Mass media law* (2009–2010 ed.). McGraw-Hill Higher Education.

Petty, R. E., Priester, J. R., & Brinol, P. (2002). Mass media attitude change: Implications of the elaboration likelihood model of persuasion. In J. Bryant & D. Zillman (Eds.), *Media effects: Advances in theory and research* (2nd ed., pp. 155–198). Lawrence Erlbaum.

Potter, W. J. (2012). *Media effects.* Sage Publications.

Price, V., Tewksbury, D., & Powers, E. (1997). Switching trains of thought: The impact of news frames on reader's cognitive responses. *Communication Research, 24*(5), 481–506. https://doi.org/10.1177/009365097024005002

Ray, M. (1973). A decision sequence analysis of developments in marketing communication. *Journal of Marketing, 37*(1), 29–38. https://doi.org/10.1177/002224297303700105

Reese, S. (2007). The framing project: A bridging model for media research revisited. *Journal of Communication, 57*(1), 148–154. https://doi.org/10.1111/j.1460-2466.2006.00334.x

Rogers, E. (1962). *Diffusion of innovations.* Free Times.

Rogers, E. (2003). *Diffusion of innovation* (5th ed.). Free Press.

Rogers, E., Dearing, J., & Bregman, D. (1993). The anatomy of agenda-setting research. *Journal of Communication, 43*(2), 68–84. https://doi.org/10.1111/j.1460-2466.1993.tb01263.x

Roskos-Ewoldsen, D. R., Roskos-Ewoldsen, B., & Carpentier, F. (2002). Media priming: A synthesis. In J. Bryant & D. Zillmann (Eds.), *Media effects: Advances in theory and research* (2nd ed., pp. 97–120). Lawrence Erlbaum.

Scheufele, D. (2013). Communicating science in social settings. *Proceedings of the National Academy of Science of the United States of America (PNAS), 110*(3), 14040–14047. https://doi.org/10.1073/pnas.1213275110

Scheufele, D., & Iyengar, S. (2014). The state of framing research: A call for new directions. In K. Kenski & K. H. Jamieson (Eds.), *The Oxford handbook of political communication theories* (pp. 619–632). Oxford University Press. https://doi.org/10.1093/oxfordhb/9780199793471.013.47

Scheufele, D. A. (1999). Framing as a theory of media effects. *Journal of Communication, 49*(1), 103–122. https://doi.org/10.1111/j.1460-2466.1999.tb02784.x

Scheufele, D. A. (2000). Agenda-setting, priming, and framing revisited: Another look at cognitive effects of political communication. *Mass Communication and Society, 3*(2&3), 297–316. https://doi.org/10.1207/S15327825MCS0323_07

Scheufele, D. A., & Tewksbury, D. (2007). Framing, agenda-setting, and priming: The evolution of three media effects models. *Journal of Communication, 57*(1), 9–20. https://doi.org/10.1111/j.0021-9916.2007.00326.x

Sheafer, T. (2007). How to evaluate it: The role of story-evaluative tone in agenda setting and framing. *Journal of Communication, 57*(1), 21–39. https://doi.org/10.1111/j.1460-2466.2006.00327.x

Sheafer, T., & Weimann, G. (2005). Agenda building, agenda setting, priming, individual voting intentions, and the aggregate results: An analysis of four Israel elections. *Journal of Communication, 55*(2), 347–365. https://doi.org/10.1111/j.1460-2466.2005.tb02676.x

Shen, F. (2004). Effects of news frames and schemas on individuals' issue interpretations and attitudes. *Journalism and Mass Communication Quarterly, 81*(2), 400–416. https://doi.org/10.1177/107769900408100211

Shen, F., & Edwards, H. H. (2005). Economic individualism, humanitarianism, and welfare reform: A value-based account of framing effects. *Journal of Communication, 55*(4), 795–809. https://doi.org/10.1111/j.1460-2466.2005.tb03023.x

Smith, E. (2002). *Energy, the environment, and public opinion*. Rowman & Littlefield Publishing.

Soroka, S. N. (2002). Issue attributes and agenda-setting by media, the public, and policymakers in Canada. *International Journal of Public Opinion Research, 14*(3), 264–285. https://doi.org/10.1093/ijpor/14.3.264

Tewksbury, D., & Scheufele, D. (2009). News framing theory and research. In J. Bryant & D. Zillmann (Eds.), *Media effects: Advances in theory and research* (3rd ed., pp. 17–33). Routledge.

Trumbo, C. (1995). Longitudinal modeling of public issues: An application of the agenda-setting process to the issue of global warming. *Journalism and Mass Communication Monographs, 152*, 1–57. https://www.proquest.com/openview/6732f6aba9b0c3ff0126ba7452934f5b/1?pq-origsite=gscholar&cbl=1818570

Tuchman, G. (1978). *Making news: A study in the construction of reality*. The Free Press.

Tuchman, G. (1993). Realism and romance: The study of media effects. *Journal of Communication, 43*(4), 36–41. https://doi.org/10.1111/j.1460-2466.1993.tb01302.x

Tversky, A., & Kahneman, D. (1981). The framing of decisions and the psychology of choice. *Science, 211*(4481), 453–458. https://doi.org/10.1126/science.7455683

United Nations. (2014). The water-energy-food nexus: A new approach in support of food security and sustainable agriculture. *FAO.org*. Retrieved March 4, 2025, from http://www.fao.org/3/a-bl496e.pdf

Vaughn, R. (1986). How advertising works: A planning model revisited. *Journal of Advertising Research, 26*(1), 57–63. https://doi.org/10.4135/9781446260807

Weaver, D. (2007). Thoughts on agenda-setting, framing, and priming. *Journal of Communication, 57*(1), 142–147. https://doi.org/10.1111/j.1460-2466.2006.00333.x

Weaver, D., McCombs, M., & Shaw, D. (2004). Agenda-setting research: Issues, attributes, and influences. In L. L. Kaid (Ed.), *Handbook of political communication research* (pp. 257–282). Lawrence Erlbaum.

Weilbacher, W. M. (2001). Point of view: Does advertising cause a 'hierarchy of effects'? *Journal of Advertising Research, 41*(6), 19–26. https://doi.org/10.2501/JAR-41-6-19-26

Xu, W., Sun, G., Lin, Z., Chen, M., Yang, B., Chen, H., & Cao, K. (2010). Knowledge, attitude, and behavior in patients with atrial fibrillation undergoing radiofrequency catheter ablation. *Journal of Interventional Cardiac Electrophysiology, 28*(3), 199–207. https://doi.org/10.1007/s10840-010-9496-2

Yao, Q. (2007). China's official framing of religion and its influence on young Chinese students: A partial testing of the process model of framing in a special media environment. *Asian Journal of Communication, 17*(4), 416–432. https://doi.org/10.1080/01292980701637033

Yao, Q. (2008). Exploring the social dynamics in the U.S. democracy: Presidential and public opinions about, and media coverage of, environmental issues. Doctoral dissertation, University of South Carolina.

Yao, Q. (2021). Informing, implying or directing? Testing the effects of message sidedness, conclusiveness and their interaction in national and local Google Ads campaigns. *Journal of Research in Interactive Marketing, 15*(4), 623–640. https://doi.org/10.1108/JRIM-09-2019-0141

Yao, Q., Joshi, P., Chang, C.-F., McDonalds, C., Tran, J., Wheeler, W., & Hou, S. (2018). Advocating a new approach to governing water, energy, and food security: Testing the effects of message inoculation and conclusion explicitness in the case of the WEF nexus. *Journal of International Crisis and Risk Communication Research, 1*(1), 111–133. https://jicrcr.com/index.php/jicrcr/article/view/6/6

Yao, Q., Liu, Z., & Stephens, L. F. (2020). Exploring the dynamics in the environmental discourse: The longitudinal interaction among public opinion, presidential opinion, media coverage, policymaking in 3 decades and an integrated model of media effects. *Environmental Systems and Decisions, 40*(1), 14–28. https://doi.org/10.1007/s10669-019-09746-y

Zaller, J. R. (1992). *The nature and origins of mass opinion*. Cambridge University Press.

CHAPTER 6

Message Factors: The Secrets to Promoting the WEF Nexus

Well-crafted persuasion messages can be powerful, as is evidenced by the old saying: "the pen is mightier than the sword" (Gee, 2015), or by the four lines from a prose poem authored by ancient Chinese writer Chi Qiu: "In March, the late spring, the grass is lush in the South of the Yangtze River; blossoms mix on the trees, and vibrant orioles fly over." This message of environmental aesthetic value purposely fulfilled its task of persuading General Bozhi Chen with his eight thousand soldiers, who lost a rebellion and fled to the North of the Yangtze River, to surrender and return to their beautiful homeland in the South of the River (Chen, 2011). The power of well-crafted persuasion messages, of course, is also exhibited in the social marketing cases discussed in Chap. 3 and the community-based social marketing (CBSM) model (McKenzie-Mohr et al., 2012). Such persuasive messages, nevertheless, can only be crafted based on a thorough understanding of persuasive techniques of message construction. Researchers, therefore, have studied hard to find the secrets of persuasion creation, which can be used to promote environmental topics such as the water-energy-food (WEF) nexus.

Q. J. Yao, *Marketing the Water-Energy-Food (WEF) Nexus*,
https://doi.org/10.1007/978-3-031-93148-2_6

1 An Overview of the Persuasiveness of Message Factors

The influence of message factors on persuasiveness has been studied intensively. Certain ways of crafting messages or incorporating factors in copywriting can produce distinct persuasive effects, as copywriters and researchers have found; popular among those factors are fear, humor, sexuality, gain- or loss-framing, and message sidedness and conclusiveness (Armstrong, 2010; O'Keefe, 2016).

The effectiveness of the fear appeal, almost the most popular persuasion factor for environmental messages on topics such as climate change, can elicit engagement and behaviors directly from the audience (Griese et al., 2018). Fear's persuasiveness is also seen in Lyndon Johnson's 1964 presidential campaign commercial, "Daisy," in which, to highlight his antinuclear stance, a little girl's face was shown with a voiceover counting from ten to zero. At the point of counting to zero, the little girl's pupil dilated, reflecting a rising mushroom cloud. This ad was broadcast only once two months before the election, and it helped President Johnson to win the election by a large margin. Fear appeals, however, must also be appropriate. Research has found that when fear stimuli in the ads increase from small to medium, desired audience responses increase from small to medium. However, beyond a certain point, when fear stimuli continue to increase, audience responses will gradually decrease (Belch & Belch, 2014). Audience responses to fear stimuli also depend on the perceived likelihood of danger and the audience's ability to overcome it. So, Armstrong (2010) suggests that persuaders use fears of dangers that can be mitigated with reasonable effort as stimuli.

Fear appeals are often seen in strategic messages on climate change or other environmental persuasion. A study notes that when people who are exposed to climate-change messages perceive high threats and high efficacy to tackle the threats, they will be more motivated by persuasion (Li & Huang, 2020). Another study finds that when people are exposed to persuasion messages portraying climate change as a low-fear issue, they feel more fearful and therefore more likely to engage in preventative or protective behaviors, particularly when they also feel morally obligated to deal with this type of issue; when climate change is portrayed as a high-fear issue, however, people will engage in preventative or protective behaviors only if they perceive a high-level collective efficacy to tackle the issue (Chen, 2016). Researchers also find that ads portraying climate change as a global threat, compared with portraying it as a local one, can obtain more audience attention, generate more positive attitudes, and change behavioral intentions more (Lee et al., 2017). Fear appeal used in some green ads, for example, promoting LED light bulbs, however, is found to generate a negative attitude toward both the ads and the products but no influence on purchase intention (Shin et al., 2017), perhaps due to the insufficient research and understanding of applying fear in this type of promotion.

Humor is also a popular persuasive appeal, which, when conveying environmental narratives, can reduce the reactance to persuasion and increase the audience's friendly behaviors toward the environment (Moyer-Guse et al., 2019). According to Armstrong (2010), 20% of TV and 30% of radio commercials as well as 10% of magazine ads use humor. A study of hundreds of college students in the U.S., Germany, and China finds that humorous ads for environmental protection generate more positive attitudes and engagements than those without humor (Griese et al., 2018). Humor appeals can capture audience attention, make people laugh, delay the irritation and wear-out that they may feel during the replay of the persuasive messages, and make those messages easy for them to remember. Research, however, has also found that humor appeals work better when used in low-involvement (small or low-priced) topics/products/brands that are already known by the audience, especially when those topics/products are perceived as fun by the audience (e.g., bottled water vs. beer). When promoting high-involvement (large or high-priced) topics/products/brands, only directly related humor should be used, moderately, as too much can interfere with the conveying of the product information. Humor should not be offensive to the audience or other people, nor should it be used in messages with serious topics, or it can be counterproductive (Armstrong, 2010).

Sexuality has also been a noticeable persuasion appeal in environmental or other campaigns. Some communicators are fascinated with using sexuality to capture the audience's eyes, which is a major reason to tarnish the reputation of the advertising industry with low taste. Critics often believe that sexual appeals insult women by portraying them as playthings. Research has also found that ads with sexual appeals attract 63% of males but only 28% of females. At the same time, 58% of females and 29% of males complain that there are too many sexual innuendos in the ads (Belch & Belch, 2014). A marketing research company, MediaAnalyzer, finds through an experiment that males fixate on the sexual elements of the ads for a longer time but pay less attention to the brand information. Ads with sexuality, compared with ads for the same products without sexuality, decrease brand recall by about half for both males and females. Therefore, Armstrong (2010) suggests that sexual appeals should only be used when they are genuinely related to topics/products/services, such as when selling body massage oil, which can reduce the resentment of females; unrelated sexual appeals should not be used so that people's attention and absorption of the advertising information will not be interfered with. Strategic environmental communicators should also bear this in mind to use sexuality effectively. Toyota, for instance, used "enough sex appeal for Hollywood celebrities" to turn its new model of Prius, an eco-friendly car that, some believe, is designed for "an environmentalist," into a hot target in the North American market (Gross, 2004).

Recently, researchers have focused more on three message factors: message sidedness, conclusiveness, and gain- or loss-framing. Their persuasive effectiveness has been examined but is still unclear, particularly in environmental communication. They will be discussed and tested in the following sections.

2 The Persuasiveness of Message Sidedness: One-Sided vs. Two-Sided Versions

In persuasive messages, is it more effective to provide only one-sided information on the positive points about the topic, or is it better to mention the negative points also and provide two-sided information? In recent years, many studies have focused on this topic, a line of research that began with Yale psychologist Carl Hovland and his colleagues (Hovland et al., 1949). In their series of studies of social psychology conducted about WWII, films of war propaganda with both the positive and negative sides were more effective to change the attitude of the audience with a pre-existing opposing attitude, and films with just the positive side were more effective to change the attitude of the audience already supporting or being less educated.

2.1 The Message Sidedness Research

Later research shows that two-sided messages with both positive and negative points are perceived as more credible than one-sided messages with only positive information (Crowley & Hoyer, 1994; Eisend, 2006; Kamins et al., 1989; O'Keefe, 2016; Pechmann, 1992). A consensus on which type is more effective in changing audience attitudes or behaviors, however, has not been reached. Some researchers find that when the opposing point is negatively related to the positive point and the opposing point is important (but less important than the positive point), two-sided persuasion is more effective in changing the audience's attitude than when the opposing point is unrelated to the positive point, or the opposing point is trivial (Pechmann, 1992). Others, however, find that two-sided persuasion is less effective when the opposing point is important to the audience (Crowley & Hoyer, 1994).

Researchers of the inoculation theory (McGuire & Papageorgis, 1961; Niederdeppe et al., 2014; Szabo & Pfau, 2002) have further found a condition for two-sided persuasion to be more convincing: refuting the negative opinions in the messages. Refutational two-sided messages have a particularly good effect when promoting a highly involved topic or when the audience suspects the credibility of the message or communicator.

The mixed findings coming out of the individual studies are confirmed by meta-analyses in this area. Allen (1991) reviews 67 experiments and concludes that, when the opposing point is raised and refuted, two-sided persuasion is more effective than its one-sided counterpart. When the opposing point is raised but not refuted, however, two-sided persuasion is less effective. Eisend (2006) reviews 34 studies and concludes that two-sided persuasion, in general, is slightly more effective than the one-sided one. O'Keefe's (2016) meta-analysis identifies two moderators (with refutation or not, being advertising or not) and finds that two-sided messages with a rebuttal to the negative points, when not perceived as advertising, are the most convincing among all the messages. This review also concludes that moderators mentioned in previous research (audience predisposition, education, involvement, etc.) make no difference.

Sidedness research has also been conducted with environmental issues or the WEF nexus. A study finds that message sidedness interacts with using humor or not to persuade farmers to get rid of the unsustainable use of anthelmintic drugs: specifically, two-sided ads with humor increase farmers' pro-environmental behavioral intention compared with their one-sided counterparts, but two-sided ads with no humor evoke negative thoughts and lead to less change in behavioral intention compared with one-sided messages (Velde et al., 2017). Another study finds that the one-sided satire from *The Onion* that claims climate change to be a hoax engages more viewers in message elaboration (critical-thinking engaging) and counterarguing (evaluation eliciting), while the two-sided satire from *The Weather Channel* that makes jokes about those who believe in human involvement in climate change is discounted as entertainment instead of a serious persuasion (Becker & Anderson, 2019). A study also suggests that culture plays a role: while the two-sided messages about climate change seemingly increase U.S. college students' knowledge, attitude, and behavioral intention on the topic, possibly because of the debating social context in America, Chinese students' knowledge, attitude, and behavioral intention toward climate change are more enhanced by one-sided messages, which could be attributed to the authoritative social environment in China (Yao et al., 2014). Another study, with 524 subjects randomly assigned into five groups for an experiment with messages about the WEF nexus and opposing points refuted in the two-sided messages, finds no difference in audience knowledge, attitude, and behavioral intention toward the WEF nexus between the two-sided groups and the one-sided groups (Yao et al., 2018). A recent study, with four field experiments using Google Ads on the topic of energy and environment, the WEF nexus, and a Higher Ed program, shows that two-sided Google ads garner more impressions, clicks, and a higher click-through rate than the one-sided Google ads in global campaigns, but one-sided Google ads generate more impressions, clicks, and a higher click-through rate with the topic of a higher education program in a local campaign, possibly because the subjects have prior knowledge of the program (Yao, 2021).

So, based on the literature reviewed above, it is hypothesized (H8): *In general, two-sided persuasion, particularly when the opposing point is refuted, is more effective than the one-sided counterpart in changing the audience's responses.*

2.2 *Testing the Message Sidedness on the WEF Nexus*

H8 is tested with the second round of the National College Student Science Literacy Survey (NCSSLS) data, with the two-sided and one-sided versions of the WEF-nexus message embedded and the opposing point refuted in the two-sided versions (see Chap. 4).

Statistical Results of the Message Sidedness Effects on the WEF Nexus
To conduct this analysis, the four behavioral groups exposed to one-sided behavioral (gain-framed, one-sided, closed-ended; gain-framed, one-sided,

open-ended; loss-framed, one-sided, closed-ended; loss-framed, one-sided, open-ended) versions were all coded into the one-sided group. The four groups exposed to the two-sided behavioral (gain-framed, two-sided, closed-ended; gain-framed, two-sided, open-ended; loss-framed, two-sided, closed-ended; loss-framed, two-sided, open-ended) versions were coded into the two-sided group. The group exposed to no message was taken as the control group. ANOVA models were conducted first to examine the differences across the three groups overall at each level of audience responses (subjective/objective knowledge, attitude, and behavioral intention) toward the WEF nexus and behavioral intention toward environmental protection. Then, ANCOVA models were used to control necessary covariates and further explore the group difference at the levels of audience responses.

Effects of Message Sidedness on Subjective WEF-Nexus Knowledge

The results of the one-way ANOVA tests (conducted through SPSS' univariate General Linear Model procedure) did not show much difference in the persuasion effects between the one-sided message group and the two-sided group, albeit some difference was suggested between the message groups and the control group (see Fig. 6.1). After being exposed to the stimulus, the one-sided group's mean perception of being well informed about the WEF nexus ($M = 3.59$, $SD = 1.53$, $N = 132$) was higher than the two-sided group ($M = 3.54$, $SD = 1.40$, $N = 123$), which was higher than the control group ($M = 3.06$, $SD = 1.53$, $N = 31$. See Table 6.1). But the p-value was 0.198, not considered significant at a conventionally accepted level. Contrasts embedded in the ANOVA test showed, however, that the mean response of the one-sided group might be considered higher than the control group ($p = 0.075$) with statistical evidence acceptable in social science studies sometimes (at an alpha level of 0.10 or 0.15 instead of the typically accepted 0.05. Iyengar, 1991; Iyengar & Kinder, 1987). The post hoc LSD comparisons reported the p-value for the difference between the two-sided group and the control group as 0.113, which was a very weak suggestion considering that LSD did not adjust for multiple comparisons. Even in that case, the p-value for the LSD post hoc test between the two-sided group and the one-sided group was still 0.769, showing no significant difference to support H8, perhaps because the single-message stimulus was not strong enough and the survey experiment was conducted in a natural setting, which provided more generalizability but reduced its capacity to capture variations (Iyengar & Kinder, 1987).

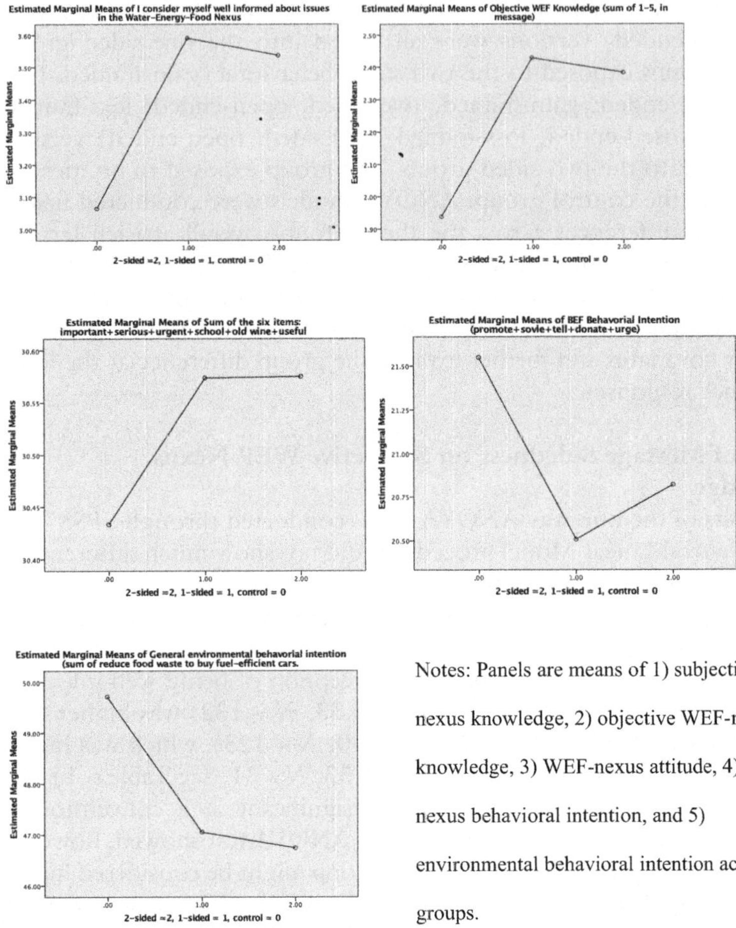

Notes: Panels are means of 1) subjective WEF-nexus knowledge, 2) objective WEF-nexus knowledge, 3) WEF-nexus attitude, 4) WEF-nexus behavioral intention, and 5) environmental behavioral intention across groups.

Fig. 6.1 Comparison of audience responses among the two-sided, one-sided, and control groups

The results turned clearer when the ANCOVA model (conducted with SPSS' univariate General Linear Model procedure) controlled possible confounding variables, including subjective WEF-nexus knowledge 1 (appearing before the message stimulus and serving as a pre-test) and family annual income. The difference in subjective WEF-nexus

Table 6.1 Comparison of audience responses among the two-sided, one-sided, and control groups

Group	*Subjective knowledge*	*Objective knowledge*	*Attitude*	*Behavioral intention*	*Environmental behavioral intention*
Two-sided	3.54 (1.40, 123)	2.38 (1.30, 128)	30.58 (5.16, 120)	20.82 (5.31, 122)	46.75 (13.29, 122)
One-sided	3.60 (1.53, 132)	2.43 (1.43, 138)	30.57 (5.42, 129)	20.51 (5.37, 132)	47.05 (13.55, 130)
Control	3.06 (1.53, 31)	1.94 (1.24, 32)	30.43 (5.52, 30)	21.55 (5.09, 31)	49.71 (11.57, 28)
Total	3.51 (1.48, 286)	2.36 (1.36, 298)	30.56 (5.30, 279)	20.75 (5.30, 285)	47.19 (13.24, 280)

Entries are the means of responses to the WEF-nexus message versions or control condition (standard deviations, sample sizes)

Homogeneous subset tests were not significant

knowledge (measured after the message stimulus) across the three groups became statistically significant ($p = 0.017$). Both the contrast test and the post hoc LSD comparison test confirmed that the mean of the one-sided group ($M = 3.80$, $SE = 0.16$) was higher than that of the control group (M = 2.78, SE = 0.31, $p = 0.004$), and the post hoc LSD test also confirmed that the mean of the two-sided group ($M = 3.56$, $SE = 0.15$) was higher than that of the control group ($p = 0.026$). The difference between the one-sided group and the two-sided group, however, was still not statistically significant ($p = 0.216$) to support H8, although the higher statistical mean of the one-sided group compared to the two-sided group is still worth noting, which suggests that two-sided versions, by mentioning the negative points of the topic, may create knowledge confusion. Clearly, however, both versions increased the self-perceived knowledge of the topic.

Also noticeable was subjective WEF-nexus knowledge 1, which positively predicted the subjective WEF-nexus knowledge ($B = 0.25$, $SE = 0.05$, $p < 0.001$), signaling the robust relationship between those differently worded measures of subjective WEF-nexus knowledge.

Effects of Message Sidedness on Objective WEF-Nexus Knowledge

The ANOVA model for objective WEF-nexus knowledge, again, reported a higher mean score in the one-sided group (M = 2.43, SD = 1.43, N = 138) than in the two-sided group (M = 2.38, SD = 1.30, N = 128), which was higher than the control group (M = 1.94, SD = 1.24, N = 32). But the overall p-value of 0.177 also made this difference hardly generalizable. The contrast test, however, provided a p-value of 0.066 (arguably acceptable in social science studies, Iyengar, 1991) for the test between the one-sided group and the control group. The post hoc LSD group comparison test provided a p-value of 0.097 for the test between the two-sided group and the control group. The p-value for the test between the one-sided group and the two-sided group was 0.788, again rendering H8 not supported.

The ANCOVA p-value for the difference across the three groups' objective WEF-nexus knowledge became statistically significant (p = 0.016) when the subjective WEF-nexus knowledge 1 and the science literacy test score were controlled. Both the contrast test and the post hoc LSD comparison test showed that the mean of the one-sided group (M = 2.48, SE = 0.11) was higher than that of the control group (M = 1.77, SE = 0.22, p = 0.004). The post hoc LSD test also showed a statistically significant difference in objective WEF-nexus knowledge between the two-sided group (M = 2.36, SE = 0.11) and the control group (p = 0.017). The difference between the one-sided group and the two-sided group, however, was still statistically insignificant (p = 0.434 with the post hoc LSD test) to support H8. Even with that, the pattern that the one-sided group scored higher in mean than the two-sided group, a pattern consistent with the subjective WEF-nexus knowledge analysis, suggests one-sidedness' effectiveness in increasing knowledge.

Also, interestingly, both the subjective WEF-nexus knowledge 1 (B = 0.11, SE = 0.04, p = 0.005) and science literacy (B = 0.29, SE = 0.05, $p < 0.001$) positively predicted the score in this objective WEF-nexus knowledge test. Subjects' self-perception of their knowledge of the WEF nexus and their science literacy are both positively correlated to their objective knowledge on this topic.

Effects of Message Sidedness on WEF-Nexus Attitude

In this ANOVA model, the difference in the attitudinal responses across the three groups was even vaguer at this higher level in the five-level model: the mean attitudes from the one-sided group (*M* = 30.57, *SD* = 5.42, *N* = 129) and the two-sided group (*M* = 30.58, *SD* = 5.16, *N* = 120) were about the same, both slightly higher than that of the control group (*M* = 30.43, *SD* = 5.52, *N* = 30), but the *p*-value (0.991) was beyond any acceptable level. The contrast test did not provide acceptable *p*-values (*p* = 0.897 for the one-sided group vs. the control group and *p* = 0.922 for the two-sided group vs. the one-sided group), and the same did the post hoc LSD comparisons. So H8 was not supported.

The ANCOVA model controlled annual family income, education, subjective WEF-nexus knowledge 1, keeping up with science news, ideology (a higher value for being more liberal), attitude toward science & technology, information consumption (print, broadcasting, social media, and talking to people), being a Democrat (= 1, else = 0), being a Protestant (= 1, else = 0), religiosity, answer quality, social trust (in a typical person in society), and objective WEF-nexus knowledge. The difference in the WEF-nexus attitude across the three groups was noticeable (*p* = 0.074) and sometimes acceptable in social science studies (Iyengar, 1991; Iyengar & Kinder, 1987). The contrast and post hoc LSD group comparison tests showed that the mean WEF-nexus attitude for the two-sided group (*M* = 30.36, *SE* = 0.60) was lower than that of the one-sided group (*M* = 32.17, *SE* = 0.61, *p* = 0.039), which was slightly lower than the control group (*M* = 32.57, *SE* = 1.29, *p* = 0.78; the *p*-value for the difference between the two-sided group and the control group was 0.128), suggesting the existence of audience psychological reactance to the message stimulus. This result reversed H8 and showed that when multiple confounders were controlled, two-sided persuasion that mentioned the criticism of the WEF nexus could actually lower the audience's supporting attitude toward the concept, even though the criticism was refuted.

Noticeable covariates included keeping up with science news (*B* = 0.52, *SE* = 0.22, *p* = 0.018), information consumption through broadcasting (*B* = 0.63, *SE* = 0.21, *p* = 0.003), and objective WEF-nexus knowledge (*B* = 0.71, *SE* = 0.34, *p* = 0.04). They all positively predicted the increase of the WEF-nexus attitude. The lack of statistically significant contribution of the subjective WEF-nexus knowledge in this ANCOVA model denoted that the audience's objective knowledge, instead of their self-perceived knowledge, built the foundation for attitude change.

Effects of Message Sidedness on WEF-Nexus Behavioral Intention

The ANOVA model for WEF-nexus behavioral intention reported little difference across the one-sided group (*M* = 20.51, *SD* = 5.37, *N* = 132), the two-sided group (*M* = 20.82, *SD* = 5.31, *N* = 30), and the control group (*M* = 21.55, *SD* = 5.09, *N* = 31). Although it appeared that the WEF-nexus behavioral intention was marginally lowered by the versions, one-sided or two-sided (compared to the control group), possibly due to the audience's psychological reactance (Yao, 2007), the *p*-value (= 0.608) made the result hardly generalizable. Neither the contrast test nor post hoc LSD tests provided a meaningful clue about the difference across the groups. H8 was again not supported.

The ANCOVA controlled gender (male = 1 and else = 0), annual family income, keeping up with science news, science literacy, information consumption with social media, being a republican (= 1 and else = 0), subjective WEF-nexus knowledge, and WEF-nexus attitude. The *p*-value for the difference across the three groups was reduced to 0.126. Meanwhile, the contrast and the post hoc LSD tests showed that the mean score for the one-sided group (*M* = 20.47, *SE* = 0.42) was lower than that of the control group (*M* = 22.41, *SE* = 0.85) with statistical significance ($p = 0.045$) and that the two-sided group's mean score (*M* = 21.07, *SE* = 0.40) was also lower than that of the control group ($p = 0.153$) but higher than the one-sided group (albeit with a *p*-value = 0.31 and rendering no statistically significant evidence for H8). Those results suggested that both groups, in fact, lowered the audience's behavioral intention toward the WEF nexus, possibly again due to the audience's psychological reactance to persuasion. The one-sided group's mean decrease was more evident than that of the two-sided group, which can serve as indirect evidence for H8.

Noticeable covariates included annual family income (*B* = −0.29, *SE* = 0.11, $p = 0.009$), keeping up with science news (*B* = 0.32, *SE* = 0.14, $p = 0.018$), using social media for information on current events (*B* = 0.29, *SE* = 0.12, $p = 0.02$), subjective WEF-nexus knowledge (*B* = 1.44, *SE* = 0.19, $p < 0.001$), and WEF-nexus attitude (*B* = 0.45, *SE* = 0.05, $p < 0.001$). While higher annual family income was negatively associated with supportive WEF-nexus behavioral intention, the latter four all positively predicted it.

Effects of Message Sidedness on Environmental Behavioral Intention

Environmental protection behavioral intention, the ANOVA reported, had almost no difference across the one-sided group (M = 47.05, SD = 13.55, N = 130), the two-sided group (M = 46.75, SD = 13.29, N = 122), and the control group (M = 49.71, SD = 11.57, N = 28, p = 0.559), similar to the pattern of the WEF-nexus behavioral intention and not supporting H8. Albeit the mean for the control group was higher than those of the groups exposed to a version of the message, hinting again at audience reactance to persuasion, the difference was hardly generalizable with p-values from either the contrast test or the post hoc LSD group comparisons.

The ANCOVA controlled age, annual family income, education, GPA, ideology, religiosity, keeping up with science news, attitude toward science and technology, social trust (in a typical person in society), frequency of talking to people for information on current events, subjective WEF-nexus knowledge, objective WEF-nexus knowledge, WEF-nexus attitude, and WEF-nexus behavioral intention. The p-value for the difference across the three groups was reduced to 0.173. The contrast and post hoc LSD tests showed that the mean of the two-sided group (M = 46.91, SE = 1.28) was lower than the one-sided group (M = 49.85, SE = 1.34, p = 0.12; reversing H8 again if a lenient alpha level was adopted as in some social science studies, e.g., Iyengar, 1991; Iyengar & Kinder, 1987), which was then lower than the mean of the control group (M = 51.30, SE = 2.65, p = 0.63; the p-value for the difference between the two-sided group and the control group was 0.14). Two-sided versions that mentioned the negative points of the WEF nexus and refuted it, when interacting with other audience characteristics, seemingly generated more audience reactance in environmental behavioral intention compared with their one-sided counterparts.

WEF-nexus behavioral intention, as a covariate, positively predicted environmental behavioral intention (B = 0.80, SE = 0.22, $p < 0.001$), the same as subjective WEF-nexus knowledge (B = 1.69, SE = 0.71, p = 0.019). In addition, Age (B = 0.41, SE = 0.15, p = 0.009) and GPA (B = 2.44, SE = 1.12, p = 0.031) also positively predicted environmental behavioral intention: older subjects or those with higher GPAs were more likely to reduce food or energy waste to protect the environment.

Interpreting the Effects of Message Sidedness on the WEF Nexus

These H8 tests examined how two-sided versions (mentioning both the positive points and negative points on the WEF nexus with the negative points refuted) and one-sided versions (mentioning only the positive points on the WEF nexus), compared with the control condition with no message stimulus, generate different persuasion effects at the response level of WEF-nexus knowledge (both subjective and objective), attitude, and behavioral intention as well as environmental behavioral intention. Given that this is a one-message-stimulus survey experiment conducted in a natural setting, although the new topic of the WEF nexus should carry little predisposition, the one-way ANOVA identified no statistically significant differences in the persuasion effects at all levels of responses. However, with a more relaxed alpha level that is sometimes used in social science studies (Iyengar, 1991; Iyengar & Kinder, 1987), the contrast and post hoc LSD comparison tests suggest that one-sided versions are more effective than the control group in generating subjective WEF-nexus knowledge, making the subjects feel more informed about the topic, which seemingly reversed H8. Both message versions seem more effective in increasing the subjects' objective WEF-nexus knowledge, enabling them to answer correctly more questions drawn from the message information, compared with the control group, although one-sided versions still appear more advantageous. Few such noticeable differences can be identified at the levels of WEF-nexus attitude, behavioral intention, and environmental behavioral intention. Thus, those results confirm that the immediate audience responses to persuasion happen at the knowledge level, and the one-sided versions are better for this task with their clearer information.

When the possibly confounding demographic and socioeconomic characteristics of the audience are controlled, some further patterns are revealed. At the knowledge levels, both the one-sided versions and the two-sided versions clearly increase the audience's subjective and objective knowledge, and the advantage of the one-sided versions over the two-sided versions is still suggested. At the attitude or behavioral level, however, persuasion seemingly creates resistance and lowers the supportive attitudes and behavioral intentions toward the topic of the WEF nexus and even the behavioral intention toward environmental protection in general. In those situations, the two-sidedness' disadvantages compared with the one-sidedness is confirmed with statistical significance at the attitudinal level, clearly reversing H8, although the one-sided versions lower the audience's WEF-nexus behavioral intention more clearly. So, H8 finds its suggestive support at the level of WEF-nexus behavioral intention when confounders are controlled and the audience's psychological reactance appears, but its opposite was supported clearly at the WEF-nexus attitude level (also in the existence of the audience resistance to persuasion) and possibly at the knowledge levels.

These findings suggest strategic environmental communicators assess their communication goals when designing a campaign to promote a new environmental topic, such as the WEF nexus. If the goal is to spread the information and educate the audience on the topic, both versions should be fine but one-sided versions are better possibly for carrying less confusion. If the goal is to change attitudes toward a new environmental topic, one-sided versions are recommended at the beginning for not raising the negative associations. Behavioral goals may not be appropriate for campaigns toward an audience with little knowledge of the topic because they can create psychological reactance in the audience. However, if behavioral intention (or behavior) goals are established, two-sided versions with the opposing point refuted may cautiously be the better choice, because, perhaps due to their perceived higher credibility (Crowley & Hoyer, 1994; Eisend, 2006; Kamins et al., 1989; O'Keefe, 2016; Pechmann, 1992), they can reduce the audience's behavioral intention less when facing the audience's psychological resistance toward persuasion.

3 The Persuasiveness of Message Conclusiveness: Conclusive vs. Inconclusive Versions

"Provide a conclusion" is traditional wisdom for communicators (Lucas, 2020; Redman & Maples, 2017), but it is challenged by the message conclusiveness research that finds, in certain circumstances, the conclusions are better left for the audience to draw instead of being stated explicitly in the persuasive messages (Armstrong, 2010; Belch & Belch, 2014; O'Keefe, 2016; Yao, 2021; Yao et al., 2014; Yao et al., 2018).

3.1 The Message Conclusiveness Research

The message conclusiveness research was also initiated by Hovland in the same series of his social psychological studies (Hovland & Mandell, 1952). The research aims to identify the more effective approach between messages with a clear (explicit) conclusion (closed-ended) and those with no clear (implicit) conclusion (open-ended). So, conclusiveness is also termed standpoint explicitness, conclusion articulation, or conclusion specificity in the literature (O'Keefe, 2016). With the war-propaganda films, Hovland's Yale research group (1952) found that conclusive persuasion changed the attitude of as many as over twice the subjects than inconclusive persuasion, but to the direction opposite to the one desired by the persuaders. That is to say, audience reactance to persuasion weakened the persuasive effects of conclusive messages more than inconclusive messages, in this discussion of WWII, a highly debated topic (Dunn, 2013). Meanwhile, several factors were found to interfere, including the audience's intellectual capacity, characteristics, and confidence in the persuader.

Some later research confirms that persuasion with explicit conclusions, although presenting clear information and carrying less confusion, can be perceived as threatening or even insulting by the audience and generate audience reactance, therefore being less convincing. On the other hand, persuasion with implicit conclusions, although risking being unclear, may be perceived as respectful, inviting, and more convincing (Martin et al., 2003; O'Keefe, 2016). Other researchers, however, find that the effects of message conclusiveness can be moderated by such variables as the audience's education (intelligence), preexisting attitude, involvement in the topic, and the complexity of the persuasion topic. For instance, for an audience with a high degree of education or involvement, or in the case of possible psychological reactance, providing only information and leaving the conclusion for the audience to make may be more effective. Otherwise, it is advisable to give clear conclusions to avoid audience confusion, as many ads do, which often even add a call to action beyond the conclusion to provide directions (Armstrong, 2010; Belch & Belch, 2014).

Even two meta-analyses of conclusiveness research yield mixed findings. O'Keefe (2016) finds little evidence to support the traditional conclusion that persuasion with implicit conclusions can be more convincing. The studies he reviewed show that inconclusive messages indeed bring in confusion, misunderstanding, and distorted perception, and therefore are less persuasive than conclusive messages. Armstrong (2010), however, believes that when the intent of persuasion is perceived by the audience and reactance is elicited, like in the case of viewing ads, persuasion with implicit conclusions (soft sell) should be more effective than persuasion with explicit conclusions (hard sell), which restricts the audience's freedom to draw their own conclusions and may push them to stick with their original position, rejecting the persuasion.

On the topic of environment, a study finds that, albeit the statistics are not significant at the commonly accepted levels (e.g., $\alpha = 0.05$), the general pattern suggests that persuasion with implicit conclusions on climate change enhances American college students' knowledge and attitude on the climate issues more than its counterpart with explicit conclusions. Messages with explicit conclusions, however, increase more Chinese college students' knowledge and attitude on the issue. An exception occurs at the behavioral level: persuasion with explicit conclusions is more effective for American college students, while persuasion with implicit conclusions is more effective for Chinese college students (Yao et al., 2014). Another study finds that the inconclusive promotional messages about the WEF nexus actually garner less audience attitude toward the WEF nexus and general behavioral intention toward environmental protection than the conclusive messages when other factors like sidedness are involved and an audience resistance to persuasion seems to appear (Yao et al., 2018). Field experiments with Google Ads find, to a global audience, the conclusive ads promoting the WEF nexus receive more impressions and clicks, although the inconclusive ads end up with a higher click-through rate, and the conclusive ads of a Higher Ed program gain more impressions in one case but lower the number of clicks in another case and have lower click-through rates in both (Yao, 2021).

These findings of the conclusiveness research, although mixed, lean toward a hypothesis that *when the topic is hotly debated and reactance to persuasion may exist, inconclusive persuasion is more convincing; but in general, or with a new topic like the WEF nexus, conclusive persuasion is more effective* (H9).

3.2 Testing Message Conclusiveness on the WEF Nexus

H9 (specifically, the second half) was also tested with the second round of the NCSSLS data.

The Statistical Results of the Message Conclusiveness Effects on the WEF Nexus

To examine the difference in the audience responses between the conclusive versions and the inconclusive versions, subjects were coded into the conclusive group, the inconclusive group, and the control group. The conclusive group consisted of the four groups exposed to the closed-ended behavioral (gain-framed one-sided, gain-framed two-sided, loss-framed one-sided, and loss-framed two-sided) versions. The inconclusive group included the four groups exposed to open-ended behavioral (gain-framed one-sided, gain-framed two-sided, loss-framed one-sided, and loss-framed two-sided) versions. The group exposed to no message served as the control group. The difference in effects was again observed at the levels of knowledge (subjective and objective), attitude, and behavioral intentions, first with the ANOVA models and then with the ANCOVA models that controlled for necessary covariates.

Effects of Message Conclusiveness on Subjective WEF-Nexus Knowledge

Regarding the difference in subjective WEF-nexus knowledge across the three groups, the ANOVA analysis showed a *p*-value ($p = 0.14$) that is occasionally considered generalizable in social science studies (Iyengar, 1991; Iyengar & Kinder, 1987). Specifically, the planned contrast to compare the mean scores of the conclusive group ($N = 121$, $M = 3.48$, $SD = 1.30$) and the inconclusive group ($N = 134$, $M = 3.64$, $SD = 1.61$) together with the mean score of the control group ($N = 31$, $M = 3.06$, $SD = 1.53$. See Table 6.2) also obtained a noticeable *p*-value ($p = 0.078$), while the planned contrast to compare the conclusive group and the

Table 6.2 Comparison of audience responses among the conclusive, inconclusive, and control groups

Group	*Subjective knowledge*	*Objective knowledge*	*Attitude*	*Behavioral intention*	*Environmental behavioral intention*
Conclusive	3.48 (1.30, 121)	2.47 (1.32, 126)	30.68 (5.14, 120)	21.12 (4.63, 122)	48.67 (12.13, 120)
Inconclusive	3.64 (1.61, 134)	2.35 (1.41, 140)	30.48 (5.44, 129)	20.23 (5.89, 132)	45.30 (14.31, 132)
Control	3.06 (1.53, 31)	1.94 (1.24, 32)	30.43 (5.52, 30)	21.55 (5.09, 31)	49.71 (11.57, 28)
Total	3.51 (1.48, 286)	2.36 (1.36, 298)	30.56 (5.30, 279)	20.75 (5.30, 285)	47.19 (13.24, 280)

Entries are the means of responses to the WEF-nexus message versions or control condition (standard deviations, sample sizes)

Homogeneous subset tests were only significant for WEF-nexus behavioral intention ($p = 0.027$) and marginal for subject WEF-nexus knowledge ($p = 0.055$)

inconclusive group was not close to statistical significance ($p = 0.38$) to support H9. The post hoc LSD group comparison test noticed that the mean score of the inconclusive group was higher than that of the control group with statistical significance ($p = 0.05$), but the *p*-value for the difference between the conclusive group and the control group was not as close ($p = 0.16$). Thus, it appeared that both groups made the subjects feel more informed about the WEF nexus, but the inconclusive versions with an open-ended conclusion did so more effectively, contrary to the prediction of H9 (see Fig. 6.2).

The ANCOVA controlled for subjective WEF-nexus knowledge 1 and annual family income (similar to the ANCOVA for sidedness on subjective WEF-nexus knowledge) and identified a statistically significant difference in the subjective WEF-nexus knowledge across the conclusive group ($M = 3.58$, $SE = 0.15$), inconclusive group ($M = 3.77$, $SE = 0.15$), and control group ($M = 2.79$, $SE = 0.31$, $p = 0.021$). The planned contrast and post hoc LSD group comparisons supported the difference between the control group and the inconclusive group with a *p*-value of 0.006, and the difference between the conclusive group and the control group with a *p*-value of 0.024. However, the *p*-value for the difference between the conclusive group and the inconclusive group was not very close to statistical significance ($p = 0.399$), still rendering no support for H9. The ANCOVA model further confirmed that both conclusive and inconclusive versions could increase subjects' self-perceived knowledge of the WEF nexus, and the inconclusive versions may be more effective (again contrary to H9), although the mean difference between the inconclusive and the conclusive versions was not statistically significant.

The ANCOVA model also showed that subjective WEF-nexus knowledge 1 (the pre-test) was positively associated with subjective WEF-nexus knowledge ($B = 0.24$, $SE = 0.05$, $p < 0.001$).

Effects of Message Conclusiveness on Objective WEF-Nexus Knowledge

For objective WEF-nexus knowledge, the *p*-value resulting from the ANOVA model for the difference across the three groups was also noticeable ($p = 0.142$). The mean scores of both the conclusive group ($N = 126$, $M = 2.47$, $SD = 1.32$) and the inconclusive group ($N = 140$, $M = 2.35$, $SD = 1.41$) were higher than the control group ($N = 32$, $M = 1.94$, $SD = 1.24$, $p = 0.064$), but the difference between the conclusive group and the inconclusive group was rather blurry ($p = 0.478$) and did not

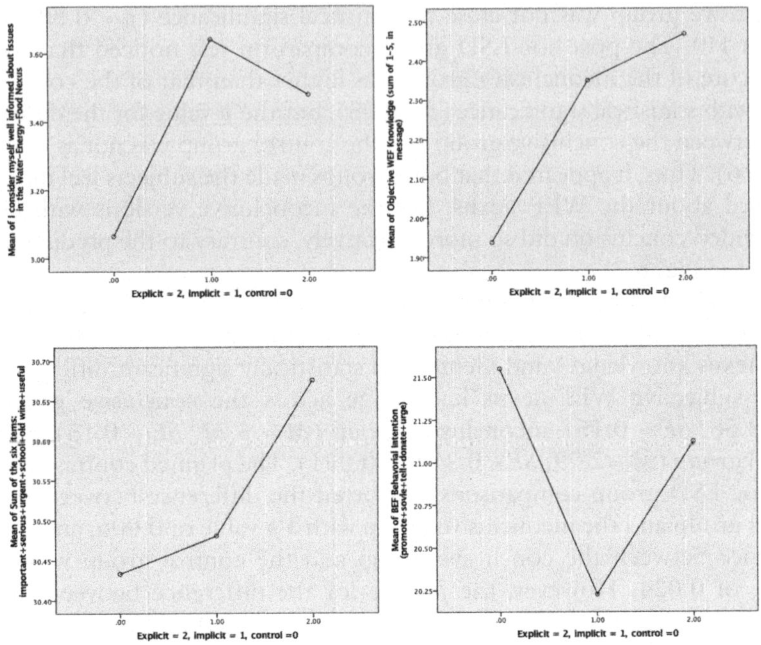

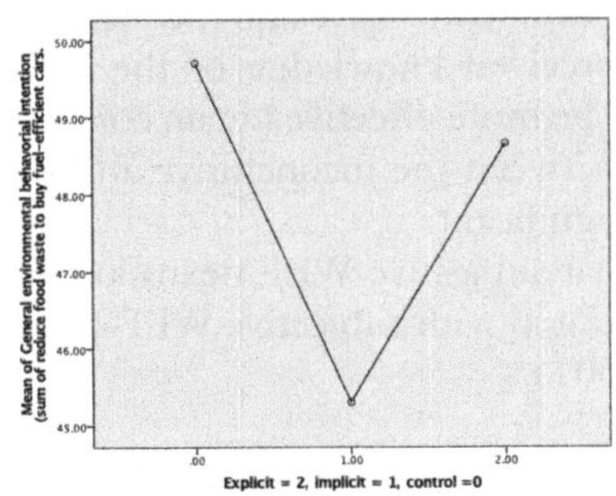

Notes: Panels are means of 1) subjective WEF-nexus knowledge, 2) objective WEF-nexus knowledge, 3) WEF-nexus attitude, 4) WEF-nexus behavioral intention, and 5) environmental behavioral intention across groups.

Fig. 6.2 Comparison of audience responses among the conclusive, inconclusive, and control groups

support H9, according to the planned contrast test. The post hoc LSD group comparison showed that the difference between the inconclusive group and the control group was noticeable ($p = 0.12$), while the difference between the conclusive group and the control group was statistically significant ($p = 0.049$). Hence, both message groups appeared to increase subjects' objective knowledge of the WEF nexus, and the increase of the conclusive versions was backed by evidence with widely accepted statistical significance, suggestively supporting H9.

The ANCOVA model controlled subjective WEF-nexus knowledge 1, annual family income, and science literacy. It reduced the *p*-value for the difference across the conclusive group ($M = 2.58$, $SE = 0.13$), the inconclusive group ($M = 2.36$, $SE = 0.13$), and the control group ($M = 1.64$, $SE = 0.27$) to statistically significant ($p = 0.009$). Compared with the control group, both the conclusive group ($p = 0.002$) and inconclusive group ($p = 0.019$) clearly increased the subjects' objective WEF-nexus knowledge, as shown in the planned contrast and the post hoc LSD group comparisons. The *p*-value for the difference between the conclusive group and the inconclusive group was reduced to 0.231, but still not statistically significant to directly support H9.

Among the covariates, both subjective WEF-nexus knowledge 1 ($B = 0.11$, $SE = 0.05$, $p = 0.018$) and science literacy ($B = 0.29$, $SE = 0.06$, $p < 0.001$) positively predicted subjects' objective WEF-nexus knowledge.

Effects of Message Conclusiveness on WEF-Nexus Attitude

According to the ANOVA model, the conclusive ($N = 120$, $M = 30.68$, $SD = 5.14$), inconclusive ($N = 129$, $M = 30.48$, $SD = 5.44$), and control ($N = 30$, $M = 30.43$, $SD = 5.52$) groups hardly made any generalizable difference at the level of WEF-nexus attitude ($p = 0.95$), although the general pattern still resembled the overall pattern of objective knowledge (see Fig. 6.2). The planned contrasts between the control group and the message groups ($p = 0.888$) and between the two message groups ($p = 0.774$, thus not supporting H9 at all) were both insignificant. According to the post hoc LSD comparisons, the differences between the conclusive group and the control group ($p = 0.82$) and between the inconclusive group and the control group ($p = 0.965$) were also insignificant. The challenge of creating attitude change through single-message persuasion was shown again in that the differences among the message groups and the control group were invisible at this attitudinal level.

The ANCOVA model lowered the *p*-value for the difference in the WEF-nexus attitude across the conclusive group, the inconclusive group, and the control group to 0.376 when controlling education, annual family income, GPA, being a Protestant, religiosity, trust (in scientists, journalists, professors, and a typical person in society respectively), participating in the first round (= 1 and else = 0), keeping up with science news, attitude toward science and technology, science literacy, ethnicity (being White and being Black, respectively), partisanship (being a Republican and being a Democrat, respectively), ideology, subjective WEF-nexus knowledge 1, and objective WEF-nexus knowledge. According to the planned contrast and the post hoc LSD comparisons, the differences between the conclusive group ($M = 31.97$, $SE = 0.60$) and the control group ($M = 31.43$, $SE = 0.27$, $p = 0.71$), between the inconclusive group ($M = 30.71$, $SE = 0.62$) and the control group ($p = 0.62$), and between the conclusive group and inconclusive group ($p = 0.16$) were all not noticeable. Although the difference between the conclusive group and the inconclusive group was the most promising one with the direction aligned with H9, the advantage of the conclusive versions compared with the inconclusive versions in enhancing supportive attitudes toward the WEF nexus, as portrayed in Fig. 6.2, is still at best conjectural and needs further examination.

The role of several covariates in shaping the WEF-nexus attitude in this model was worth noting. Objective WEF-nexus knowledge, instead of subjective WEF-nexus knowledge, was positively associated with WEF-nexus attitude ($B = 0.93$, $SE = 0.34$, $p = 0.007$) and served as the foundation to cultivate the latter. Trust in professors, a major source of science and environmental information (among whom the author of this book), positively predicted subjects' supportive attitude toward the WEF nexus ($B = 1.09$, $SE = 0.44$, $p = 0.015$). Also, Black subjects showed more likelihood to support the WEF nexus ($B = 4.36$, $SE = 1.67$, $p = 0.01$). Other covariates gained no statistically significant parameter estimates but were instrumental in reducing the *p*-value of the ANCOVA model.

Effects of Message Conclusiveness on WEF-Nexus Behavioral Intention

Because the test of homogeneity of variances for the three groups at the level of WEF-nexus behavioral intention was statistically significant ($p = 0.027$), which meant that the variances of the WEF-nexus behavioral intention across the three groups were not equal to begin with, the analysis of the results of the ANOVA test (p = 0.275) was complicated. The general pattern also suggested persuasion reactance at this level (see Fig. 6.2), showing that the mean score of the control group (N = 31, M = 21.55, SD = 5.09) was higher than those of the conclusive group (N = 122, M = 21.12, SD = 4.63) and the inconclusive group (N = 132, M = 20.23, SD = 5.89), although the p-value was not statistically significant (p = 0.375, with no assumption of equal variances), according to the planned contrast test. The planned contrast between the conclusive group and the inconclusive group was also not statistically noticeable (p = 0.177, with no assumption of equal variances, thus not supporting H9), nor did the difference between the conclusive group and the control group (p = 0.69) or between the inconclusive group and the control group (p = 0.213) identified by the post hoc LSD comparisons.

The ANCOVA model reported a p = 0.011 for the difference across the conclusive group (M = 21.63, SE = 0.41), inconclusive group (M = 20.14, SE = 0.43), and the control group (M = 22.50, SE = 0.87) when it controlled annual family income, science literacy, keeping up with science news, attitude toward science and technology, subjective WEF-nexus knowledge 1, subjective WEF-nexus knowledge, and WEF-nexus attitude. Again, based on the planned contrast and post hoc LSD comparisons, possibly due to the resistance to persuasion, both the conclusive (p = 0.37) and inconclusive groups (p = 0.016) lowered the subjects' supportive behavioral intention toward the WEF nexus, compared with the control group. The advantage of the conclusive versions, compared with the inconclusive versions, at the level of WEF-nexus behavioral intention, was supported with statistical significance (p = 0.013), supporting H9. Conclusive versions' straightforwardness and clarity on the WEF nexus helped to reduce the subjects' behavioral intention toward supporting the topic less than did the inconclusive versions when facing the audience's psychological reactance.

Among the covariates, annual family income played a clearer role in lowering the subjects' behavioral intention toward the WEF nexus (B = −0.31, SE = 0.11, p = 0.006), while keeping up with science news, not surprisingly, positively contributed to it (B = 0.43, SE = 0.13, p = 0.001). Subjective WEF-nexus knowledge (B = 1.41, SE = 0.20, p < 0.001), instead of the subjective WEF-nexus knowledge 1 (B = −0.11, SE = 0.17, p = 0.51) or objective WEF-nexus knowledge (excluded from the model in the process of the model-building exploration), positively predicted behavioral intention toward the WEF nexus in this model.

WEF-nexus attitude, also with no surprise, positively contributed to the increase of WEF-nexus behavioral intention (B = 0.48, SE = 0.05, $p < 0.001$), supporting the hierarchy of response models.

Effects of Message Conclusiveness on Environmental Behavioral Intention

The effects of message conclusiveness on environmental protection behavioral intention maintained the same pattern as on WEF-nexus behavioral intention (see Fig. 6.2), revealing subjects' reactance to persuasion, and the p-value ($p = 0.074$) for the difference across the three groups produced by the ANOVA model was much more noticeable (Iyengar, 1991; Iyengar & Kinder, 1987). The planned contrast did not identify a noticeable difference ($p = 0.299$) between the mean scores of the control group ($N = 28$, $M = 49.71$, $SD = 11.57$) and the message groups, but it identified a statistically significant difference between the conclusive group ($N = 120$, $M = 48.67$, $SD = 12.13$) and the inconclusive group ($N = 132$, $M = 45.30$, $SD = 14.31$, $p = 0.044$), thus supporting H9. Furthermore, the post hoc LSD comparisons showed that the difference between the inconclusive group and the control group was also statistically noticeable ($p = 0.108$. Iyengar, 1991; Iyengar & Kinder, 1987), although the difference between the conclusive group and the control group was much ignorable ($p = 0.705$). So, it is declared with certain confidence that the conclusive versions were more effective than the inconclusive versions in keeping the subjects' behavioral intention to protect the environment less reduced when both versions seemingly decreased the environmental behavioral intention as they suffered from psychological reactance to persuasion.

The ANCOVA controlled age, education, GPA, ideology, religiosity, social trust (in a typical person in society), keeping up with science news, attitude toward science and technology, subjective WEF-nexus knowledge, WEF-nexus attitude, and WEF-nexus behavioral intention. It identified a p-value of 0.038 for the difference in environmental behavioral intention across the conclusive group ($B = 49.64$, $SE = 1.16$), the inconclusive group ($B = 45.43$, $SE = 1.11$), and the control group ($B = 48.10$, $SE = 2.26$). The difference in enhancing the environmental behavioral intention between the conclusive versions and the inconclusive versions was further confirmed with evidence of statistical significance ($p = 0.011$) provided in the planned contrast and the post hoc LSD group comparisons, officially supporting H9. The difference between the conclusive group and the control group ($p = 0.295$) and the difference between the inconclusive group and the control group ($p = 0.54$) were both still not statistically significant. At this level and with these covariates controlled, it seemed that the inconclusive versions stimulated audience reactance to persuasion, but the conclusive versions, with their advantage of providing clear information and an explicit conclusion, kept the subjects' environmental behavioral intention at a level higher than both the control group and the inconclusive versions.

Covariates contributing to the subjects' environmental behavioral intention included education ($B = -1.10$, $SE = 0.55$, $p = 0.045$), GPA ($B = 1.83$, $SE = 0.90$, $p = 0.043$), ideology ($B = 3.50$, $SE = 0.91$, $p < 0.001$), WEF-nexus attitude ($B = 0.47$, $SE = 0.18$, $p = 0.008$), and WEF-nexus behavioral intention ($B = 0.81$, $SE = 0.19$, $p < 0.001$). It was not a surprise to learn that both WEF-nexus attitude and behavioral intention positively contributed to environmental behavioral intention and that the more liberal a subject's ideology was, the more likely the subject tended to have the behavioral intention to help solve the environmental issues overall. It was surprising, however, to see that when the positively contributing GPA was under control, education was negatively associated with environmental behavioral intention. Not all educated people think that the environment needs to be taken care of, which is probably why environmental issues are so tricky to tackle, and more effective promotions are needed to cultivate more support.

Interpreting the Effects of Message Conclusiveness on the WEF Nexus

The ANOVA tests overall did not provide statistically significant results at a conventionally accepted alpha level across the three groups at any level of audience response. However, the differences across the groups at the levels of subjective WEF-nexus knowledge, objective WEF-nexus knowledge, and environmental behavioral intention are still noticeable, with a more lenient alpha level that is sometimes used in social science studies (Iyengar, 1991; Iyengar & Kinder, 1987). Similar to the results of message sidedness, both the conclusive and the inconclusive versions seemingly increase subjective and objective knowledge but suffer from the audience's reactance to persuasion at the behavioral levels. The inconclusive versions make the subjects feel more informed about the WEF nexus, but the conclusive versions increase their objective knowledge and environmental behavioral intention.

The ANCOVA models, with different sets of necessary covariates under control, further provide statistically significant evidence for the pattern that both the conclusive and the inconclusive versions are effective in increasing subjective knowledge and objective knowledge but stimulate audience reactance perhaps at the attitudinal level and clearly at the level of WEF-nexus behavioral intention. Facing audience resistance to persuasion, conclusive versions are more advantageous to keep the WEF-nexus behavioral intentions less reduced compared with the inconclusive versions, suggesting that clear conclusions should be recommended for the promotion of new environmental issues such as the WEF nexus. At the level of environmental behavioral intention, the conclusive versions increase the mean score to a higher level than that of the control group, reducing resistance to persuasion, while the mean score of inconclusive versions remains lower than that of the control group.

Overall, the data seemingly support O'Keefe's (2016) conclusion while opposing those of Hovland and associates (1952) and Armstrong (2010), aligning with H9's prediction of the effectiveness of the conclusive versions of the WEF-nexus message, albeit inconclusive versions show more persuasiveness in increasing audience subjective knowledge. The suggestion to strategic communicators on new environmental topics such as the WEF nexus is to pilot test the persuasive messages first, particularly because the difference between the conclusive messages and the inconclusive messages is not very clear overall and is perhaps moderated by many factors, like those controlled in the ANCOVA models. If the time restriction does not allow a pilot campaign, conclusive persuasion is safer to choose.

4 The Persuasiveness of Gain- or Loss-Framing

"Is the glass half full or half empty?" people's answers to this question are often said to reflect their worldviews. It is not clear how many will choose "half full" or "half empty," but a study conducted by psychologists Daniel Kahneman and Amos Tversky (1979) may suggest the answer. Their study presented subjects with the opportunity to choose between two options: the first offered a 50% chance for a three-week tour and 50% for nothing, and the second offered a 100% chance for a one-week tour. Among their subjects, 78% chose the second option, and only 22% chose the first, demonstrating loss-aversion psychology as stated in their prospect theory.

4.1 The Gain- or Loss-Framing Research

The prospect theory (Kahneman & Tversky, 1979), or "loss-aversion theory," has been studied in many areas, including strategic communications, where it has evolved into the popular gain- or loss-framing research (de Alcantara et al., 2020; Ghesla et al., 2020; Jensen et al., 2018; Nabi et al., 2018; Nan et al., 2016; O'Keefe & Nan, 2012).

The gain- or loss-framing research suggests that persuasion messages should focus on avoiding the losses in lieu of acquiring the gains for better effectiveness (Jensen et al., 2018; Nabi et al., 2018), because of the audience's internal mechanism of loss-aversion. For instance, researchers find that warning labels emphasizing the risk of sugar-taking are more effective in driving subjects to choose low-sugar food products than health logos that focus on the gain of low-sugar food (de Alcantara et al., 2020).

The gain- or loss-framing research has been largely applied to promoting vaccination, a subfield of health or science communication. In a meta-analysis of 32 studies (11,814 subjects in total), O'Keefe and Nan (2012) find that in the experiments on vaccination promotion, a large number of studies have shown no difference in the persuasiveness between the messages focusing on the benefits and those focusing on the losses, although the latter have some vague advantages in persuading the parents, slightly confirming the loss-aversion assumption. If other factors are considered, the effectiveness of the gain- or loss-framing may change. For instance, researchers find that when the possibility for a child to be infected is high, ads for vaccination that focus on the benefits persuade the parents more. But when the parents feel that the child is less likely to be infected, ads for vaccination that focus on the loss are more convincing (Nan et al., 2016). The gain- or loss-framing research has also been quickly used to explain coping strategies and vaccination promotion during COVID since its outbreak in 2019, and researchers find that Chinese adults show no difference in responses to gain- or loss-framed persuasion promoting COVID vaccination (Chen et al., 2022). People in the U.S. and the Netherlands, however, seemingly respond more positively to the gain-framed persuasion to promote stricter regulations to deal with COVID than the loss-framed persuasion in a survey experiment because the gain-framed persuasion elicits support for risk-aversion strategies (Hameleers, 2021).

Several meta-analyses of gain- or loss-framing research in a broader context have also found that the evidence is complicated and little difference in the persuasiveness between the two frames can be concluded (Jensen et al., 2018). For instance, O'Keefe and Jensen (2006) analyze 165 studies and identify no difference between gain- and loss-frames to support the prospect theory; instead, gain-framed persuasion actually can be slightly more effective when promoting disease prevention behaviors. With little evidence supporting it, some scholars even believe that the occurrence of the prospect theory is likely to be a special case in the laboratory and will not happen in the real world; they, therefore, caution strategic communicators to use the gain- or loss-frame in persuasion carefully (Armstrong, 2010).

Some researchers have also called for the next generation of gain- or loss-framing research to explore possible moderators that together eliminate the difference in the persuasiveness between those two frames (Jensen et al., 2018). They have also connected the gain- or loss-framing research with the fear-appeal research (by treating the loss-frame as a fear appeal), explored moderators of the gain- or loss-frame's persuasiveness, such as self-efficacy, grit, consideration of future consequences, health information overload, which are drawn from the fear-appeal research literature, and found them functioning in some situations when promoting physical activity to different audiences (Jensen et al., 2018). Acknowledging that the extensive amount of research is still unable to explain when and why one message frame, gain or loss, can be more persuasive than the other, some researchers have lately also proposed an integrative framework, which predicts that the persuasiveness of gain- or loss-framing is moderated by whether the information processing is systematic or heuristic and whether the goal of the information processing is accuracy, defense, or impression (Nan et al., 2018). They also find that although gain- or loss-framing research has been connected with different theories such as regulatory focus theory, dual-process theories, and psychological reactance theory, prospect theory is still the dominant one to guide the gain- or loss-framing research, which lays the theoretical foundation for 89.6% of the 126 studies that they reviewed. Researchers have also found that, when audience resistance to persuasion (i.e., audience psychological reactance) happens as a moderator, gain-framed persuasion generates lower reactance, greater freedom threat, and a more positive attitude on the topic of organ donation, more positive attitude on the topic of weight loss, and no difference on the topic of safe-driving, but a less perceived threat to freedom on the topic of sun safety and skin cancer (Reynolds-Tylus, 2019).

In strategic environmental communication, the persuasiveness of the gain- or loss-framing has also been studied and generated mixed findings, particularly on climate change, which is often loss-framed through warnings (Nabi et al., 2018). A study with field and survey experiments, for instance, finds that loss-framed arguments to call to improve the climate-change situation are less mobilizing compared with their gain-framed counterparts, even when those loss-framed messages are designed with more persuasion power, possibly because the loss-framed messages remind the subjects their own personal issues and potential losses to deal with and make them less willing to spend time and energy on tackling climate change, a common issue (Levine & Kline, 2019). Another study reveals that the gain-frame of climate change evokes more hope, and the loss-frame evokes more fear, both directly or indirectly leading to increases in attitude and advocacy behaviors; it is then suggested that messages evoking fear first and then hope (such as the gain-framed messages following a threatening message) will have the strongest positive impact on audience advocacy behaviors (Nabi et al., 2018). To convert people from meat-centric diets, which are linked to public health, animal suffering, and environmental issues, to plant-based diets in a university setting, the gain-framed message that promotes plant-based meals is more effective than the loss-framed message that focuses on restricting meat-based meals in generating positive responses by both the sensory measures (e.g., make the plant-based meals tastier) and the behavioral measures (e.g., do not serve meals with meat), although the behavioral measures change not as much as the sensory measures (Carvalho et al., 2022). A field study also notices that, to promote energy-conservation consumption in households, the loss-framed persuasion (if the household does not reach the energy-saving goal it will lose a tree that will be planted) is more effective than the gain-framed persuasion (the household will be planted a tree if reaching the energy-saving goal) and persuades the households saving 5% monthly consumption compared with the control group (Ghesla et al., 2020).

Although no studies of gain- or loss-framing on the relatively new topic of the WEF nexus have been found yet, based on the extensive research of the gain- or loss-framing on other scientific or environmental issues reviewed above, it is interesting to examine whether, in the context of WEF nexus (H10): *Loss-framed messages, compared to the gain-framed messages, generate more favorable audience responses.* This is also very likely the first study of gain- or loss-framing on this specific topic of sustainability.

4.2 *Testing Gain- or Loss-Framing on the WEF Nexus*

H10 was also tested with the second round of the NCSSLS data.

Statistical Results of the Gain- or Loss-Framing Effects on the WEF Nexus

To test the possible difference between the gain-framed versions and the loss-framed versions, three groups were coded: the gain-framed (including the gain-framed, two-sided, open-ended; gain-framed, two-sided, closed-ended; gain-framed, one-sided, open-ended; and gain-framed, one-sided, closed-ended) behavioral group, the loss-framed (including the loss-framed, two-sided, open-ended; loss-framed, two-sided, closed-ended; loss-framed, one-sided, open-ended; and loss-framed, one-sided, closed-ended) behavioral group, and the control group (exposed to no message). As done in the analyses of message sidedness and conclusiveness, both ANOVA and ANCOVA were used to test the differences across the groups at each level of response.

Effects of Gain- or Loss-Framing on Subjective WEF-Nexus Knowledge

The ANOVA for the mean score of subjective WEF-nexus knowledge across the gain-framed group (N = 126, M = 3.69, SD = 1.45; see Table 6.3), the loss-framed group (N = 129, M = 3.44, SD = 1.48), and the control group (N = 31, M = 3.06, SD = 1.53, p = 0.08) was not statistically significant with a conventional level of alpha but noticeable for social scientists sometimes (Iyengar, 1991; Iyengar & Kinder, 1987; see Fig. 6.3). Based on the planned contrast and the post hoc LSD group comparisons, the difference between the gain-framed group, with the highest mean score, and the control group, with the lowest mean score, was statistically significant (p = 0.035). The difference between the gain-framed group and the loss-framed group also demands further exploration (p = 0.178), although it was not statistically significant to support H10. The difference between the loss-framed group and the control group was even vague (p = 0.21). The gain-frame has more advantages in making the subjects feel more informed about the WEF nexus than the loss-frame, contrary to H10.

Table 6.3 Comparison of audience responses among the gain-framed, loss-framed, and control groups

Group	*Subjective knowledge*	*Objective knowledge*	*Attitude*	*Behavioral intention*	*Environmental behavioral intention*
Gain-framed	3.69 (1.45, 126)	2.39 (1.30, 133)	30.63 (5.52, 122)	20.59 (5.58, 125)	47.49 (13.97, 124)
Loss-framed	3.44 (1.48, 129)	2.43 (1.44, 133)	30.52 (5.08, 127)	20.72 (5.10, 129)	46.34 (12.85, 128)
Control	3.06 (1.53, 31)	1.94 (1.24, 32)	30.43 (5.52, 30)	21.55 (5.09, 31)	49.71 (11.57, 28)
Total	3.51 (1.48, 286)	2.36 (1.36, 298)	30.56 (5.30, 279)	20.75 (5.30, 285)	47.19 (13.24, 280)

Entries are the means of responses to the WEF-nexus message versions. or control condition (standard deviations, sample sizes)

No homogeneous subsets tests were significant

The ANCOVA model controlled subjective WEF-nexus knowledge 1 and annual family income once again, resulting in $p = 0.03$ for the differences among the gain-framed group ($M = 3.69$, $SE = 0.15$), the loss-framed group ($M = 3.66$, $SE = 0.15$), and the control group ($M = 2.79$, $SE = 0.31$). The planned contrast test verified the difference between the loss-framed group and the control group ($p = 0.013$), but the difference between the loss-framed and the gain-framed groups remained unclear ($p = 0.89$), not supporting H10. The post hoc LSD group comparisons further confirmed that the difference between the gain-framed group and the control group was also statistically significant ($p = 0.011$). Again, both the gain-frame and the loss-frame increased subjective WEF-nexus knowledge when the two covariates were controlled, but the difference between the two message groups was not identifiable.

Regarding the covariates, annual family income was a vital control variable but had no statistically significant contribution to the variation of subjective WEF-nexus knowledge. Subjective WEF-nexus knowledge 1, the pre-test, positively predicted subjective WEF-nexus knowledge ($B = 0.25$, SE $= 0.05$, $p < 0.001$), as found in the analyses of the message sidedness and conclusiveness.

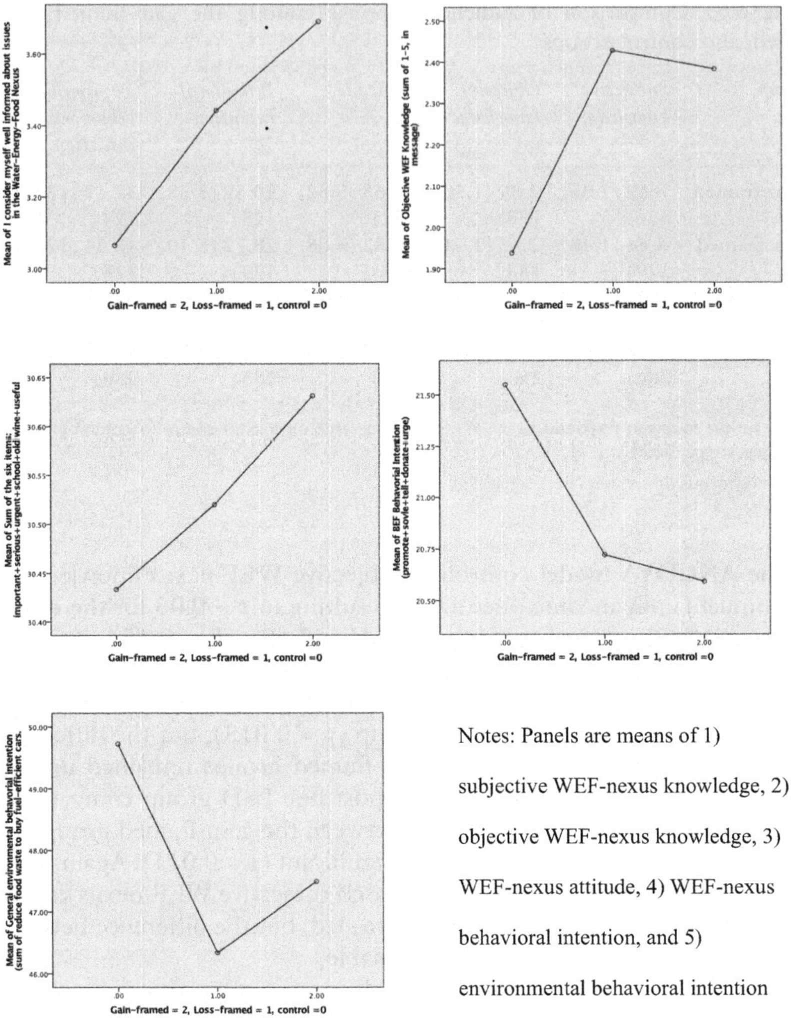

Notes: Panels are means of 1) subjective WEF-nexus knowledge, 2) objective WEF-nexus knowledge, 3) WEF-nexus attitude, 4) WEF-nexus behavioral intention, and 5) environmental behavioral intention across groups.

Fig. 6.3 Comparison of audience responses among the gain-framed, loss-framed, and control groups

Effects of Gain- or Loss-Framing on Objective WEF-Nexus Knowledge

The difference in the mean scores of the objective WEF-nexus knowledge among the gain-framed group ($N = 133$, $M = 2.38$, $SD = 1.30$), the loss-framed group ($N = 133$, $M = 2.43$, $SD = 1.44$), and the control group ($N = 32$, $M = 1.94$, $SD = 1.24$, $p = 0.177$) was indiscernible in the ANOVA model. The planned contrast showed that the difference between the message groups and the control group was noticeable ($p = 0.066$) but the difference between the gain-framed and the loss-framed groups was negligible ($p = 0.786$), lending no supporting evidence for H10. The post hoc LSD group comparison test further confirmed that the difference between the gain-framed group and the control group ($p = 0.096$) and the difference between the loss-framed group and the control group ($p = 0.067$) were both noticeable with a more tolerable level of alpha used in some social science studies (Iyengar, 1991; Iyengar & Kinder, 1987). Both frames appeared to enhance the subjects' objective WEF-nexus knowledge.

The ANCOVA model controlled subjective WEF-nexus knowledge 1, annual family income, and science literacy. A $p = 0.017$ was reported for the difference in the mean scores across the gain-framed group ($M = 2.44$, $SE = 0.13$), the loss-framed group ($M = 2.51$, $SE = 0.13$), and the control group ($M = 1.65$, $SE = 0.27$). The ANCOVA planned contrast test clearly confirmed that the loss-frame increased the objective WEF-nexus knowledge compared to the control group ($p = 0.005$), although the difference between the gain-frame and the loss-frame was still ambiguous ($p = 0.687$), not supporting H10. The post hoc LSD comparisons further showed that the gain-frame increased the objective WEF-nexus knowledge compared to the control group ($p = 0.01$). Again, the results demonstrated that either frame increased the objective knowledge, but the advantage of one frame over the other was not supported or even suggested by the data.

For the covariates, both subjective WEF-nexus knowledge 1 ($B = 0.11$, $SE = 0.05$, $p = 0.023$) and science literacy ($B = 0.29$, $SE = 0.06$, $p < 0.001$) positively contributed to objective WEF-nexus knowledge, as shown in the ANCOVA models for sidedness and conclusiveness.

Effects of Gain- or Loss-Framing on WEF-Nexus Attitude

The difference in the WEF-nexus attitude created by the gain-framed group ($N = 122$, $M = 30.63$, $SD = 5.52$), the loss-framed group ($N = 127$, $M = 30.52$, $SD = 5.08$), and the control group ($N = 30$, $M = 30.43$,

$SD = 5.52$, $p = 0.977$), was unclear in the ANOVA model. The contrast test also detected little difference between the message groups and the control group ($p = 0.89$) and between the gain-framed group and the loss-framed group ($p = 0.869$), not supporting H10 at all; so did the post hoc LSD comparisons between the gain-framed group and the control group ($p = 0.855$) and between the loss-framed group and the control group ($p = 0.936$). Although the pattern of the statistical means of WEF-nexus attitude was in the shape consistent with those of the subjective and possibly objective WEF-nexus knowledge (see Fig. 6.3), those differences were nearly impossible to generalize beyond the sample.

The ANCOVA model controlled trust (in journalists, professors, and a typical person in society, respectively), ideology (higher value standing for being more liberal), religion (being a Protestant and being a Muslim, respectively), subjective health status, keeping up with science news, information consumption (through websites and talking to people, respectively), participation in the first round, subjective WEF-nexus knowledge 1, and subjective WEF-nexus knowledge. The *p*-value for the difference in WEF-nexus attitude across the gain-framed group ($M = 30.39$, $SE = 0.46$), the loss-framed group ($M = 31.44$, $SE = 0.42$), and the control group ($M = 29.78$, $SE = 0.93$) was lowered to 0.118, noticeable in social science studies (Iyengar, 1991; Iyengar & Kinder, 1987). The contrast showed that both the difference between the gain-framed and the loss-framed groups ($p = 0.099$, marginally supporting H10) and the difference between the loss-framed group and the control group ($p = 0.104$) were noticeable and worthy of further exploration, with the loss-framed versions seemingly showing an advantage of increasing the WEF-nexus attitude. The post hoc LSD group comparisons also showed that the difference between the gain-framed group and the control group ($p = 0.564$) was very vague.

Among the covariates, trust in professors ($B = 1.37$, $SE = 0.30$, $p < 0.001$) clearly positively contributed to the WEF-nexus attitude. Being a Muslim somehow decreased WEF-nexus attitude ($B = -8.89$, $SE = 3.25$, $p = 0.007$). Using websites for information on current events increased support for the WEF nexus ($B = 0.35$, $SE = 0.16$, $p = 0.026$), but talking to family, friends, or other people for that purpose decreased it ($B = -0.36$, $SE = 0.16$, $p = 0.024$), possibly because subjects who enjoyed gaining information on current events through talking did not care environmental issues and the usage of water, energy, and food resources as much. Subjective WEF-nexus knowledge 1 (the pretest) had a clear positive contribution to WEF-nexus attitude ($B = 0.56$, $SE = 0.18$, $p = 0.002$), but subjective WEF-nexus knowledge was unrelated to it ($B = 0.11$, $SE = 0.21$, $p = 0.60$), although it still needed to be controlled to reduce the overall *p*-value of the ANCOVA model. Objective WEF-nexus knowledge also had a statistically significant and robust positive contribution to WEF-nexus attitude ($B = 0.87$, $SE = 0.23$, $p < 0.001$), but its exclusion from the ANCOVA model would bring the *p*-value of the model down from 0.207 to 0.118; so, it was not included in this model.

Effects of Gain- or Loss-Framing on WEF-Nexus Behavioral Intention

The difference in the mean score of the WEF-nexus behavioral intention detected by the ANOVA model across the gain-framed group ($N = 125$, $M = 20.59$, $SD = 5.58$), the loss-framed group ($N = 129$, $M = 20.72$, $SD = 5.10$), and the control group ($N = 31$, $M = 21.55$, $SD = 5.09$) was as vague as that at the level of the WEF-nexus attitude. The contrast test found no meaningful difference between the message groups and the control group ($p = 0.378$) and between the gain-framed group and the loss-framed group ($p = 0.847$), thus again not supporting H10 at all. The post hoc LSD group comparisons also found no significant differences between the gain-framed group and the control group ($p = 0.371$) and between the loss-framed group and the control group ($p = 0.437$). Although the shape of the statistical means of the three groups suggested that an audience reactance to persuasion was again activated; the message groups decreased the behavioral intention to support the WEF nexus compared with the control group (see Fig. 6.3), but the pattern was not substantial enough to have statistical significance.

The ANCOVA model lowered the *p*-value for the difference in WEF-nexus behavioral intention across the gain-framed group ($M = 20.34$, $SE = 0.36$), the loss-framed group ($M = 20.96$, $SE = 0.35$), and the control group ($M = 22.38$, $SE = 0.70$) to 0.037 when it controlled keeping up with science news, attitude toward science and technology, trust in professors, consuming information on current events through websites, science literacy, subjective WEF-nexus knowledge, and the WEF-nexus attitude. According to the planned contrast, the difference between the loss-framed group and the control group was noticeable ($p = 0.07$; Iyengar, 1991; Iyengar & Kinder, 1987), but the difference between the loss-framed group and the gain-framed group was equivocal ($p = 0.218$), failing to support H10. The post hoc LSD group comparison showed that the difference between the gain-framed group and the control group was statistically significant ($p = 0.011$). Again, it appeared that the audience's resistance to persuasion decreased effects on WEF-nexus behavioral intention for both message groups compared with the control group, and the gain-framed versions' lowering of WEF-nexus behavioral intention was even statistically significant compared with the control group, suggestively supporting H10's prediction of the loss-frame advantage in the context of audience reactance to persuasion.

Three covariates produced statistically significant positive contributions to WEF-nexus behavioral intention: keeping up with science news ($B = 0.31$, $SE = 0.13$, $p = 0.017$), subjective WEF-nexus knowledge ($B = 1.31$, $SE = 0.16$, $p < 0.001$), and, not surprisingly, based on the hierarchy of response model, WEF-nexus attitude ($B = 0.49$, $SE = 0.05$, $p < 0.001$). Other covariates did not have a statistically significant contribution but helped keep the p-value for the ANCOVA model lower.

Effects of Gain- or Loss-Framing on Environmental Behavioral Intention

The difference in the mean of the environmental protection behavioral intention across the gain-framed group ($N = 124$, $M = 47.49$, $SD = 13.97$), the loss-framed group ($N = 128$, $M = 46.34$, $SD = 12.85$), and the control group ($N = 28$, $M = 49.71$, $SD = 11.57$), as detected by the ANOVA model, was also not generalizable ($p = 0.447$). The planned contrast reported that the p-value for the difference between the message groups and the control group was 0.289 and that for the difference between the gain-framed group and the loss-framed group was 0.489, rendering H10 unsupported again. The post hoc LSD group comparisons reported that the difference between the gain-framed group and the control group was 0.223 and that between the loss-framed group and the control group was 0.423. So, the pattern of the effects across the three groups once more showed that reactance to persuasion decreased environmental behavioral intention and that the gain-framed versions helped to reduce that decrease. Again, without statistical significance, the difference could not be generalized.

The ANCOVA model reduced the p-value for the gain-framed group ($B = 49.69$, $SE = 1.35$), loss-framed group ($B = 46.77$, $SE = 1.31$), and the control group ($B = 51.66$, $SE = 2.67$) to a noticeable degree ($p = 0.148$; Iyengar, 1991; Iyengar & Kinder, 1987) when it controlled age, annual family income, religiosity, ideology, social trust (in a typical person in society), information consumption (through print media, social media, and talking respectively), subjective WEF-nexus knowledge 1, objective WEF-nexus knowledge, and WEF-nexus behavioral intention. The planned contrast test showed that the differences both between the loss-framed group and the control group ($p = 0.105$) and between the loss-framed group and the gain-framed group ($p = 0.126$) were noticeable, hence suggestively reversing H10. The p-value generated by the post hoc LSD group comparisons for the difference between the gain framed group and the control group was very weak ($p = 0.517$), indicating again that, facing reactance to persuasion, the gain-frame probably suffered less than the loss-frame in lowering the subjects' environmental behavioral intention.

Among the covariates, age ($B = 0.29$, $SE = 0.14$, $p = 0.035$), objective WEF-nexus knowledge ($B = 1.50$, $SE = 0.71$, $p = 0.036$), and WEF-nexus behavioral intention ($B = 1.17$, $SE = 0.17$, $p < 0.001$) showed statistically positive contribution to environmental behavioral intention. WEF-nexus behavioral intention, as expected, was the most robust one, because it was at the same level of message effects as the environmental behavioral intention.

Interpreting the Effects of Gain- or Loss-Framing on the WEF Nexus

The ANOVA models and their associated contrast and post hoc group comparison tests detected some noticeable differences across the gain-framed group, the loss-framed group, and the control group at the levels of subjective and objective WEF-nexus knowledge but not WEF-nexus attitude, WEF-nexus behavioral intention, and environmental behavioral intention. Specifically speaking, the gain-framed versions are clearly more effective in increasing the audience's subjective WEF-nexus knowledge than the control group, but the loss-framed versions fail to do so. Both frames seem to increase objective WEF-nexus knowledge (Iyengar, 1991; Iyengar & Kinder, 1987). Few generalizable differences are detectable by the ANOVA models at the attitudinal and behavioral intention levels, but their patterns are consistent with those of the persuasion effects of the sidedness and conclusiveness, showing audience reactance that makes both frames lower the audience's WEF-nexus attitude and behavioral intention as well as environmental behavioral intention, compared to the control group.

The ANCOVA models, through controlling possible confounders, reveal more insights into the persuasiveness of the gain- or loss-framing. Regarding the subjective and objective WEF-nexus knowledge, both the gain-frame and the loss-frame are clearly more effective than the control group condition in increasing the mean scores, while the effectiveness between the gain-frame and the loss-frame has no difference. At the attitudinal level, the loss-frame is arguably more effective than the gain-frame or the control condition, but the gain-frame makes no difference from the control group. At the behavioral levels, the audience reactance to persuasion happens again and makes the message groups lower the subjects' WEF-nexus and environmental behavioral intentions. With that audience reactance, loss-frame is more advantageous to lower WEF-nexus behavioral intention less and maintains an obscure difference from the control group, while the gain-frame's difference from the control group is statistically significant. The loss-frame, however, is disadvantageous in the case of the environmental behavioral intention as its difference with the control group and the gain-frame can be both detectable with a lenient level of alpha, but the difference between the gain-frame and the control group is not supported with data evidence of statistical significance.

In sum, the prospect theory (H10) seems supported at the levels of WEF-nexus attitude and behavioral intention with its assumption of the loss-aversion psychological mechanism in the audience, but some evidence supports the advantage of the gain-frame persuasion in increasing subjective knowledge of WEF nexus. It appears that, with the topic of WEF nexus, if the communication goal is to increase audience attitude or behavioral intention, loss-frame should be recommended. But if the goal is to make the audience feel more informed or encourage more general behaviors of environmental protection, such as eating less meat, saving energy, or buying eco-friendly vehicles, gain-frame may be the choice. For the goal of spreading WEF-nexus knowledge, nevertheless, the gain-frame and the loss-frame do not differ much.

5 The Interactions of the Message Factors

This is likely the first study that examines together the three message factors (sidedness, conclusiveness, and gain- or loss-frame), popular among copywriters and strategic communicators, which provides a valuable opportunity to examine whether, and if so how, they interact with each other in the persuasion process. Examining such interactions deepens our understanding of their persuasiveness for using them more effectively in strategic environmental communications.

5.1 *The Possible Interactions of Message Factors: Sidedness, Conclusiveness, and Gain- or Loss-Framing*

Some studies have analyzed and confirmed the interaction between message sidedness and message conclusiveness on the topic of the WEF nexus (Yao, 2021; Yao et al., 2018). In one study (Yao et al., 2018), data show that the inconclusive two-sided message of WEF nexus is not as effective as the conclusive one-sided message in increasing audience knowledge of WEF nexus in a test. Also, when the persuasion is two-sided, no difference is discerned between the open-ended and the closed-ended messages; but when the persuasion is one-sided, the closed-ended messages are more effective in increasing the audience's positive attitude toward the WEF nexus and general behavioral intention toward environmental protection than are the open-ended messages. Another study (Yao, 2021) finds that, in a global campaign, the two-sided open-ended ad of a higher ed program generates the highest number of clicks and click-through rate, while the one-sided closed-ended ad generates the lowest number of them, and their difference is statistically significant. When the ads are open-ended, the two-sided ads generate higher numbers of impressions, clicks, and a higher click-through rate. In a local campaign, however, the one-sided open-ended ad of the higher ed program generates the highest number of clicks and click-through rate, while the two-sided open-ended ad has the lowest number of both. Also, when the ads are one-sided, the open-ended ad generates more clicks and a higher click-through rate than the closed-ended ad; when two-sided, the closed-ended ad secures more clicks and a higher click-through rate than the open-ended ad. The interaction between message sidedness and conclusiveness is quite clear in these studies.

Research, however, is also needed to fill a gap, clarify the mixed results, and shed light on how message sidedness and conclusiveness interact with gain- or loss-framing simultaneously, which has also been found to interact with other variables (Jensen et al., 2018; Nan et al., 2018; Reynolds-Tylus, 2019). Researchers have constantly called for exploring possible moderations and interactions to bring more insights into the role of these three message factors. For instance, message-sidedness researchers have proposed humor (Eisend, 2022), education, preexisting attitude, topic complexity, audience involvement, and the magnitude of the opposing attitude as the possible moderators of the advantages of one-sided or two-sided persuasion (e.g., Bohner et al., 2003; Crowley & Hoyer, 1994;

Pechmann, 1992). Message conclusiveness researchers speculate that the moderation of such factors as education (intelligence), preexisting attitude, audience involvement, and topic complexity may help explain more of the persuasiveness of the open-ended and closed-ended messages (Ahearne et al., 2000; Armstrong, 2010; Belch & Belch, 2014; Kardes, 1988; Sawyer & Howard, 1991). Gain- or loss-framing researchers have also called for an exploration of moderators in their second generation of research, proposing possible moderators such as fear appeal, self-efficacy, grit, consideration of future consequences, health information overload (Jensen et al., 2018), information processing mode (systematic vs. heuristic), information processing goal (accuracy, defense, or impression; Nan et al., 2018), and audience resistance to persuasion, or audience psychological reactance (Reynolds-Tylus, 2019).

Hence a research question (R1): *Do interactions exist among message sidedness, conclusiveness, and the gain- or loss-frame, and if so, how do the interactions function on the persuasiveness at each level of audience response, on the topic of the WEF nexus?*

5.2 Exploring the Interactions of the Message Factors on the WEF Nexus

The second round of NCSSLS data was again used to answer this research question.

Statistical Results of Effects of the Interactions on the WEF Nexus

Sidedness, conclusiveness, and gain- or loss-framing seemingly made some differences when one message factor was combined with another, as seen from the group statistics (see Table 6.4) and the overall pattern (see Fig. 6.4) across the eight behavioral message groups and the control group. To test their possible interactions, the variables of sidedness, conclusiveness, and gain- or loss-framing were all included in the three-way ANOVAs (but the control group was coded as missing; so only eight behavioral message groups were in the analysis) with audience responses at the levels of knowledge, attitude, and behavioral intentions respectively as the response variables. Then, for further insight, a set of three-way ANCOVAs controlling different covariates was conducted to observe their interactions.

Table 6.4 Comparison of audience responses among the eight message groups and the control group

Group	*Subjective knowledge*	*Objective knowledge*	*Attitude*	*Behavioral intention*	*Environmental behavioral intention*
Gain-framed, two-sided, open-ended	3.66(1.61, 29)	2.55 (1.18, 31)	33.07 (4.70, 27)	22.68 (5.88, 28)	50.31 (14.47, 29)
Gain-framed, two-sided, closed-ended	3.27 (1.17, 30)	2.50 (1.19, 32)	29.66 (5.45, 29)	19.97 (4.67, 30)	46.10 (12.90, 30)
Gain-framed, one-sided, open-ended	4.08 (1.52, 36)	2.19 (1.43, 37)	29.09 (5.99, 35)	18.51 (6.55, 35)	44.24 (15.92, 34)
Gain-framed, one-sided, closed-ended	3.68 (1.40, 31)	2.33 (1.36, 33)	31.16 (5.11, 31)	21.63 (4.04, 32)	49.77 (11.75, 31)
Loss-framed, two-sided, open-ended	3.60 (1.54, 35)	2.14 (1.38, 36)	29.26 (5.06, 35)	19.71 (5.58, 35)	42.94 (13.87, 35)
Loss-framed, two-sided, closed-ended	3.62 (1.24, 29)	2.38 (1.45, 29)	30.76 (4.79, 29)	21.24 (4.68, 29)	48.50 (10.85, 28)
Loss-framed, one-sided, open-ended	3.21 (1.70, 34)	2.56 (1.59, 36	31.16 (5.12, 32)	20.50 (4.96, 34)	44.53 (12.43, 34)
Loss-framed, one-sided, closed-ended	3.35 (1.38, 31)	2.66 (1.31, 32)	31.06 (5.30, 31)	21.61 (5.11, 31)	50.19 (12.97, 31)
Control	3.06 (1.53, 31)	1.94 (1.24, 32)	30.43 (5.52, 30)	21.55 (5.09, 31)	49.71 (11.57, 28)
Total	3.51 (1.48, 286)	2.36 (1.36, 298)	30.56 (5.30, 279)	20.75 (5.30, 285)	47.19 (13.24, 280)

Entries are the means of response to the WEF-nexus message versions or control condition (standard deviations, sample sizes)

No homogeneous subsets tests were significant

Effects of the Interactions on Subjective WEF-Nexus Knowledge

The three-way ANOVA of sidedness, conclusiveness, and gain- or loss-framing at the level of subjective WEF-nexus knowledge showed a significant interaction between sidedness and gain- or loss-framing ($p = 0.042$). While the interaction between conclusiveness and gain- or loss-framing

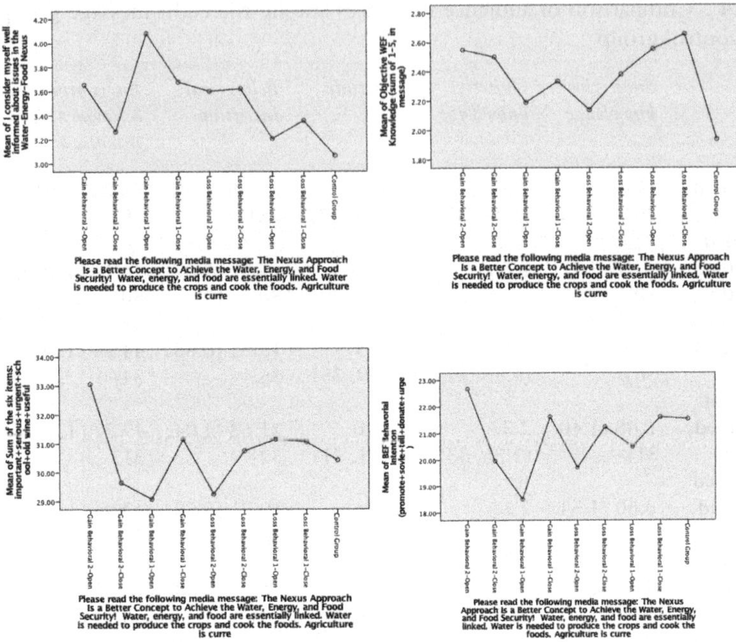

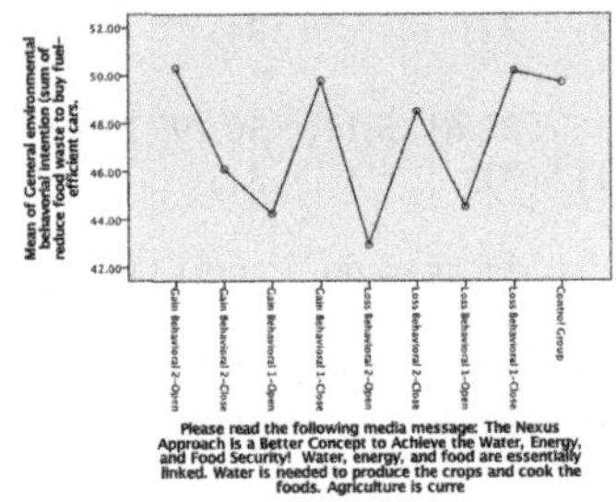

Notes: Panels are means of 1) subjective WEF-nexus knowledge, 2) objective WEF-nexus knowledge, 3) WEF-nexus attitude, 4) WEF-nexus behavioral intention, and 5) environmental behavioral intention across groups.

Fig. 6.4 Comparison of audience response among the eight message and control groups

was also arguably noticeable ($p = 0.191$; Iyengar, 1991, Iyengar & Kinder, 1987), the main effects of the three factors and the other interactions were ignorable. For the interaction between sidedness and gain- or loss-framing, specifically, with gain-frame, two-sidedness ($N = 59$, $M = 3.46$, $SD = 1.41$)

was less effective than one-sidedness (N = 67, M = 3.90, SD = 1.47) in making the audience feel more informed about the WEF nexus; but with loss-frame, two-sidedness (N = 64, M = 3.61, SD = 1.40) became more effective than one-sidedness (N = 65, M = 3.28, SD = 1.55). Seemingly, when consuming information on gains, one-sidedness provided clear information and were effective, while two-sidedness might have created confusion. But when consuming information on losses, two-sidedness provided points from both sides to consider the trade-offs, while one-sidedness, which did not provide the negative points, was considered less informative (see the pattern in Fig. 6.5). Overall, the gain-framed one-sided versions generated the highest mean score in this interaction, and the loss-framed one-sided versions generated the lowest.

For the marginally significant interaction between conclusiveness and gain- or loss-framing, when the versions were gain-framed, inconclusive versions (N = 65, M = 3.89, SD = 1.56) made the audience feel more informed about the WEF nexus than the conclusive versions (N = 61, M = 3.48, SD = 1.30); when the versions were loss-framed, conclusive versions (N = 60, M = 3.48, SD = 1.31) became more effective than the inconclusive versions (N = 69, M = 3.41, SD = 1.62; see Fig. 6.5 for the pattern). The gain-framed inconclusive versions were the most effective in this interaction to generate the highest mean score, and the loss-framed inconclusive versions generated the lowest mean score among the four groups, but further exploration would be needed to generalize this conclusion, given its relatively large p-value.

The three-way ANCOVA model controlled subjective WEF-nexus knowledge 1 and information consumption through websites and talking, respectively, lowered the p-value for the interaction between sidedness and gain- or loss-framing to as low as 0.008, but none of the p-values for the main effects or other interactions were close to statistical significance. For the interaction between sidedness and gain- or loss-framing, again, with gain-frame, one-sidedness (N = 65, M = 3.89, SD = 1.49) made the audience feel more informed about the WEF nexus than two-sidedness (N = 58, M = 3.43, SD = 1.40); with loss-frame, however, two-sidedness (N = 63, M = 3.63, SD = 1.39) was more effective than one-sidedness (N = 65, M = 3.28, SD = 1.55), which confirmed the findings of the three-way ANOVA model discussed above. Again, the gain-framed one-sided versions generated the highest mean score, and the loss-framed one-sided versions had the lowest mean.

Panel 1: at the level of subjective WEF-nexus knowledge

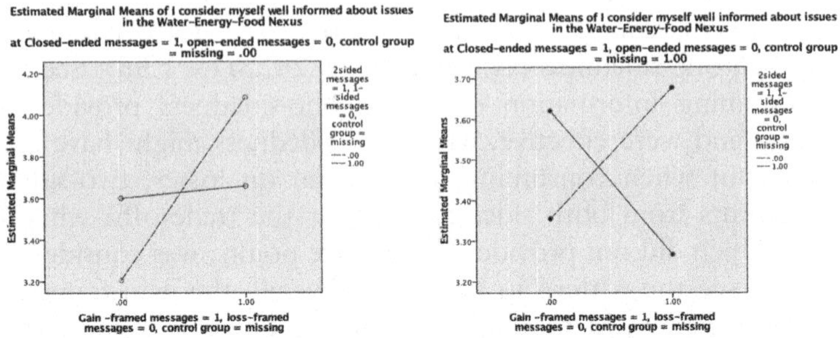

Panel 2: at the level of objective WEF-nexus knowledge

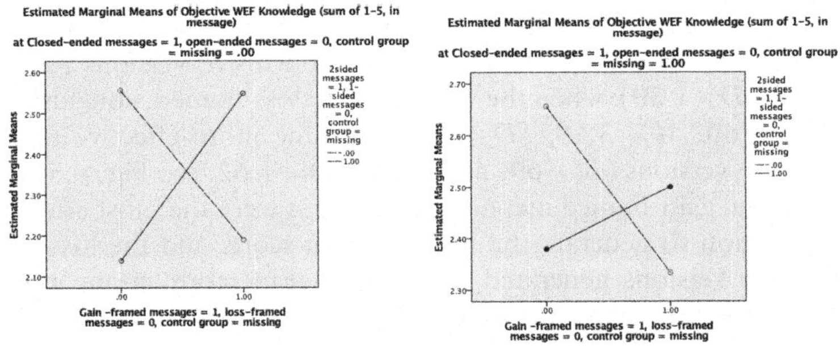

Panel 3: at the level of WEF-nexus attitude

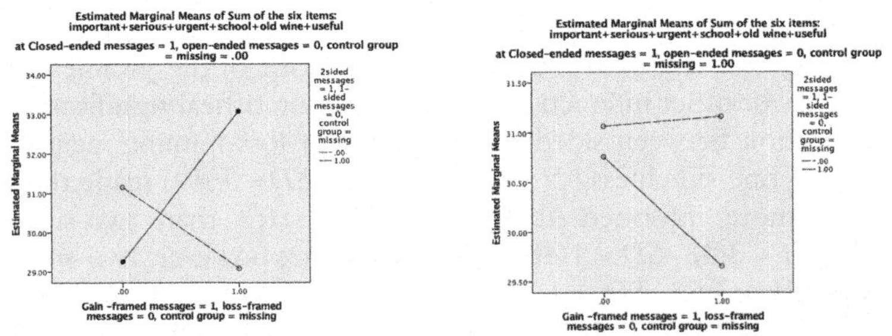

Fig. 6.5 The interactions among sidedness, conclusiveness, and gain- or loss-framing

Panel 4: at the level of WEF-nexus behavioral intention

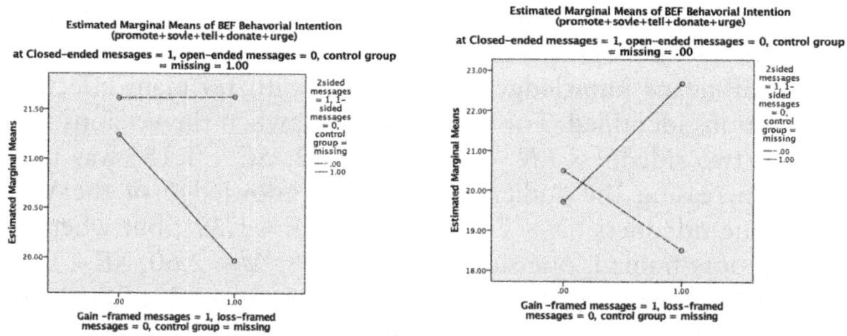

Panel 5: at the level of environmental behavioral intention across groups

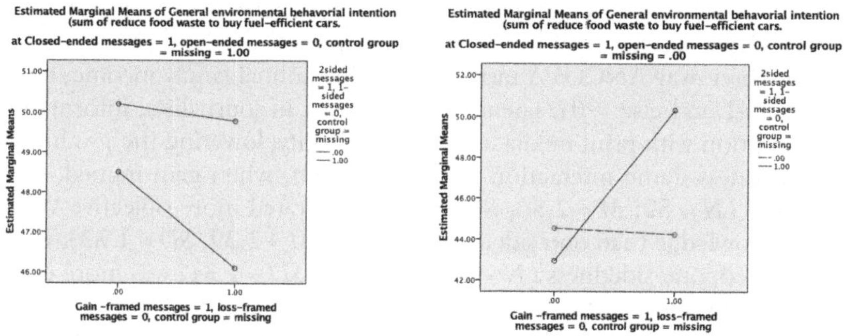

Fig. 6.5 (continued)

For the covariates, subjective WEF-nexus knowledge 1 ($B = 0.24$, $SE = 0.05$, $p < 0.001$) and information consumption through websites ($B = 0.11$, $SE = 0.05$, $p = 0.016$) were both found to positively contribute to subjective WEF-nexus knowledge, but information consumption through talking to people ($B = -0.12$, $SE = 0.05$, $p = 0.015$) negatively contributed to it. So, in this context, gaining information on current events through talking to family, friends, or other people made the audience feel less informed about the WEF nexus.

Effects of the Interactions on Objective WEF-Nexus Knowledge

This three-way ANOVA again showed a noticeable *p*-value for the interaction between sidedness and gain- or loss-framing contributing to the objective WEF-nexus knowledge ($p = 0.072$), with no main effects or other interactions identified. For this interaction, when the versions were gain-framed, two-sidedness ($N = 63$, $M = 2.52$, $SE = 1.18$) was more effective in increasing the audience's objective knowledge of the WEF nexus than one-sidedness ($N = 70$, $M = 2.26$, $SE = 1.39$); but when the versions were loss-framed, one-sidedness ($N = 68$, $M = 2.60$, $SE = 1.46$) was more effective than the two-sidedness ($N = 65$, $M = 2.25$, $SE = 1.40$). The loss-framed one-sided versions arguably increased the mean score of the objective knowledge test the most, while the loss-framed two-sided versions increased the least. The pattern of interaction was opposite to the interaction between the same two variables at the level of subjective knowledge (of course, this *p*-value is less convincing).

The three-way ANCOVA model controlled annual family income, being White (= 1 and else = 0), science literacy, trust in journalists, information consumption with print media, and answer quality, lowering the *p*-value for the sidedness-frame interaction to 0.017. Again, when gain-framed, two-sidedness ($N = 52$, $M = 2.56$, $SD = 1.21$) generated more objective WEF-nexus knowledge than one-sidedness ($N = 45$, $M = 2.29$, $SD = 1.53$); when loss-framed, one-sidedness ($N = 45$, $M = 2.71$, $SD = 1.34$) was more effective in increasing objective knowledge of the WEF nexus than two-sidedness ($N = 46$, $M = 2.24$, $SD = 1.42$). The loss-framed one-sided versions were still the best performer among these four groups, and the loss-framed two-sided versions performed again the worst. This result further confirmed that the persuasive approach may differ significantly between increasing subjective and objective WEF-nexus knowledge.

The three-way ANCOVA model also lowered the *p*-value for the three-way interaction to almost statistically significant ($p = 0.052$). The three-way interaction meant that, to increase the audience's objective WEF-nexus knowledge, for gain-frame, two-sidedness worked better with implicit conclusions ($N = 25$, $M = 2.68$, $SD = 1.22$) than with explicit conclusions ($N = 27$, $M = 2.44$, $SD = 1.22$), while one-sidedness worked better with explicit conclusions ($N = 22$, $M = 2.50$, $SD = 1.50$) than with implicit conclusions ($N = 23$, $M = 2.09$, $SD = 1.56$); for loss-frame, on the other hand, two-sidedness worked better with explicit conclusions ($N = 22$, $M = 2.45$, $SD = 1.40$) than with implicit conclusions ($N = 24$, $M = 2.04$, $SD = 1.43$), while one-sidedness worked better with implicit conclusions

($N = 19$, $M = 2.74$, $SD = 1.41$) than with explicit conclusions ($N = 26$, $M = 2.69$, $SD = 1.32$). While the loss-framed, one-sided, inconclusive version was the best performer among these eight groups, the loss-framed, two-sided, inconclusive version performed the worst. This three-way interaction demonstrated the complexity of persuasion effects even at a single level of audience response.

For the covariates, science literacy ($B = 0.26$, $SE = 0.05$, $p < 0.001$), answer quality ($B = 0.36$, $SE = 0.05$, $p < 0.001$), and being White ($B = 0.38$, $SE = 0.19$, $p = 0.045$) all positively contributed to objective WEF-nexus knowledge, but using print media to acquire information on current events, surprisingly, was negatively associated with it ($B = -0.14$, $SE = 0.04$, $p < 0.001$).

Effects of the Interactions on WEF-Nexus Attitude

The three-way ANOVA to analyze the contribution of the three message factors to WEF-nexus attitude revealed a noticeable interaction between sidedness and gain- or loss-frame again ($p = 0.079$), a possible one between sidedness and conclusiveness ($p = 0.143$), and clearly a statistically significant one among all three factors ($p = 0.008$). The significant three-way interaction was apparently the most meaningful finding because it guided the interpretations of two-way interactions and simple main effects in itself (Jaccard & Turrisi, 2003). For the interaction between sidedness and gain- or loss-framing, gain-frame increased WEF-nexus attitude more when two-sided ($N = 56$, $M = 31.30$, $SD = 5.34$) than one-sided ($N = 66$, $M = 30.06$, $SD = 5.65$), and loss-frame was more effective when one-sided ($N = 63$, $M = 31.11$, $SD = 5.17$) than two-sided ($N = 64$, $M = 29.94$, $SD = 4.96$). The top performer was the gain-framed two-sided versions, and the bottom performer was the loss-framed two-sided versions. This pattern (see Fig. 6.5) was somewhat opposite to the pattern of the same interaction at the level of subjective WEF-nexus knowledge, and the reason needed further exploration. For the interaction between sidedness and conclusiveness, it appeared that one-sidedness was more effective when conclusive ($N = 62$, $M = 31.11$, $SD = 5.16$) than inconclusive ($N = 67$, $M = 30.07$, $SD = 5.64$); two-sidedness was more effective when inconclusive ($N = 62$, $M = 30.92$, $SD = 5.23$) than conclusive ($N = 58$, $M = 30.21$, $SD = 5.11$). The one-sided conclusive versions had the highest mean score, and the two-sided conclusive versions had the lowest. The three-way interaction, with robust statistical significance, showed that for gain-frame, two-sidedness worked more effectively being inconclusive ($N = 27$, $M = 33.07$, $SD = 4.70$) than conclusive ($N = 29$, $M = 29.66$, $SD = 5.45$), and one-sidedness was more effective being conclusive (N = 31, M = 31.16, SD = 5.11) than inconclusive (N = 35, M = 29.09, SD = 5.99); for loss-frame, however, two-sidedness was more effective to increase WEF-nexus attitude when conclusive (N = 29, M = 30.76, SD = 4.79) than inconclusive (N = 35, M = 29.26, SD = 5.06), and one-sidedness was more effective when inconclusive (N = 32, M = 31.16, SD = 5.12) than conclusive (N = 31, M = 31.06, SD = 5.30). Among these eight groups, the gain-framed, two-sided, inconclusive version had the highest mean score, and the gain-framed, one-sided, inconclusive version had the lowest. This three-way interaction was consistent in pattern with the three-way interaction revealed by the three-way ANCOVA at the level of objective WEF-nexus knowledge, except that the best and the worst performers were different.

The three-way ANCOVA for WEF-nexus attitude controlled first-round participation, gender (male = 1 and else = 0), trust in journalists and professors respectively, being Black (= 1 and else = 0), being a Muslim (= 1 and else = 0), information consumption through broadcasting, subjective WEF-nexus knowledge, and objective WEF-nexus knowledge. It reduced the p-value for the interaction between sidedness and gain- or loss-frame to be statistically significant ($p = 0.037$). The p-value for the three-way interaction was no longer statistically significant at the conventional level of alpha (= 0.05) but still noticeable ($p = 0.057$). For the interaction between sidedness and gain- or loss-frame, clearly, gain-frame was more persuasive in producing WEF-nexus attitude when two-sided (N = 56, M = 31.30, SD = 5.34) than one-sided (N = 65, M = 29.95, SD = 5.62); loss-frame generated more WEF-nexus attitude when one-sided (N = 63, M = 31.11, SD = 5.17) than two-sided (N = 64, M = 29.94, SD = 4.96). The best performer was the gain-framed two-sided group, and the worst was the loss-framed two-sided group, although the difference between the four groups was not very much. The three-way interaction confirmed that gain-framed two-sided versions were more effective when inconclusive (N = 27, M = 33.07, SD = 4.70) than conclusive (N = 29, M = 29.66, SD = 5.45), and gain-framed one-sided versions were more effective being conclusive (N = 30, M = 30.97, SD = 5.08) than inconclusive (N = 35, M = 29.09, SD = 5.99); on the other hand, loss-framed two-sided versions were more effective to increase WEF-nexus attitude when conclusive (N = 29, M = 30.76, SD = 4.79) than inconclusive (N = 35, M = 29.26, SD = 5.06), and loss-framed one-sided versions were more effective when inconclusive (N = 32, M = 31.16, SD = 5.12) than conclusive (N = 31, M = 31.06, SD = 5.30). Among these eight groups, the best performer was the gain-framed, two-sided, inconclusive version, and the worst was the gain-framed, one-sided, inconclusive version.

Seven covariates positively contributed to WEF-nexus attitude: trust in journalists (B = 0.48, SE = 0.24, p = 0.044) and professors (B = 1.14, SE = 0.27, p < 0.001), being Black (B = 2.77, SE = 0.97, p = 0.005), consuming information on current events through broadcasting (B = 0.29, SE = 0.12, p = 0.017), first-round participation (B = 1.43, SE = 0.62, p = 0.021), and objective WEF-nexus knowledge (B = 1.08, SE = 0.22, p < 0.001). The positive contribution of subjective WEF-nexus knowledge (B = 0.38, SE = 0.20, p = 0.059) was also noticeable. Meanwhile, being male (B = −1.31, SE = 0.58, p = 0.025) or a Muslim (B = −6.90, SE = 2.75, p = 0.013), in this context, reduced WEF-nexus attitude, which was consistent with the previous finding that, in America, males were less enthusiastic about environmental topics (Yao et al., 2014).

Effects of the Interactions on WEF-Nexus Behavioral Intention

At the level of the WEF-nexus behavioral intention, the three-way ANOVA model identified a statistically significant two-way interaction between sidedness and conclusiveness (p = 0.042) and, more meaningfully and robustly, again a three-way interaction among all three message factors (p = 0.019). For the two-way interaction, to generate more WEF-nexus behavioral intention, two-sidedness needed to be inconclusive (N = 63, M = 21.03, SD = 5.86) instead of conclusive (N = 59, M = 20.59, SD = 4.69), and one-sidedness worked better being conclusive (N = 63, M = 21.62, SD = 4.56) than inconclusive (N = 69, M = 19.49, SD = 5.86), a pattern mostly consistent with the probable interaction between sidedness and conclusiveness at the level of WEF-nexus attitude. The best performer of these four groups was the one-sided conclusive group, and the worst was the one-sided inconclusive group. In the three-way interaction, for gain-frame, to be more effective, two-sidedness needed to be inconclusive (N = 28, M = 22.68, SD = 5.88) than conclusive (N = 30, M = 19.97, SD = 4.69), and one-sidedness needed to be conclusive (N = 32, M = 21.63, SD = 4.04) instead of inconclusive (N = 35, M = 18.51, SD = 6.55); for loss-frame, two-sidedness needed to be conclusive (N = 29, M = 21.24, SD = 4.68) than inconclusive (N = 35, M = 19.71, SD = 5.58), and the same happened to one-sidedness (being conclusive, N = 31, M = 21.61, SD = 5.11; being inconclusive, N = 34, M = 20.50, SD = 4.96; see Fig. 6.2). In this three-way interaction, loss-frame performed differently than those in the three-way interaction at the level of WEF-nexus attitude. The group with the highest mean score was exposed to the gain-framed, two-sided, inconclusive version, and the lowest-scored group was exposed to the gain-framed, one-sided, inconclusive version.

The three-way ANCOVA model controlled annual family income, information consumption through print media and social media respectively, subjective WEF-nexus knowledge, and WEF-nexus attitude revealed robust statistical significance in the main effect of conclusiveness ($p = 0.008$) and, more meaningfully in this case, the two-way interaction between sidedness and conclusiveness ($p = 0.01$). The two-way interaction between sidedness and gain- or loss-framing was also noticeable ($p = 0.10$). For the main effect of the conclusiveness, clearly, conclusive versions ($N = 93$, $M = 21.32$, $SD = 4.83$) generated more WEF-nexus behavioral intention than the inconclusive versions ($N = 87$, $M = 20.40$, $SD = 6.24$). For the two-way interaction between sidedness and conclusiveness, two-sidedness generated more WEF-nexus behavioral intention when inconclusive ($N = 46$, $M = 21.48$, $SD = 6.11$) than conclusive ($N = 47$, $M = 20.72$, $SD = 4.92$); one-sidedness was more effective when conclusive ($N = 46$, $M = 21.93$, $SD = 4.72$) than inconclusive ($N = 41$, $M = 19.20$, $SD = 6.23$). Among these four groups, the group exposed to the one-sided conclusive versions had the highest mean score of WEF-nexus behavioral intention, while the group exposed to the one-sided inconclusive versions had the lowest. For the probable two-way interaction between sidedness and gain- or loss-framing, gain-frame generated more WEF-nexus behavioral intention when two-sided ($N = 48$, $M = 21.40$, $SD = 5.87$) than one-sided ($N = 42$, $M = 20.02$, $SD = 5.77$), and loss-frame generated more when one-sided ($N = 45$, $M = 21.22$, $SD = 5.48$) than two-sided ($N = 45$, $M = 20.78$, $SD = 5.18$). The best-performing group was exposed to the gain-framed two-sided versions, and the worst was exposed to the gain-framed one-sided versions, although the mean scores for the four groups had not much difference. The p-value for this interaction was also only arguably noticeable in social science studies (Iyengar, 1991; Iyengar & Kinder, 1987).

Among the covariates, information consumption with print media ($B = 0.35$, $SE = 0.14$, $p = 0.011$) and social media ($B = 0.34$, $SE = 0.13$, $p = 0.01$) for current events, subjective WEF-nexus knowledge ($B = 1.49$, $SE = 0.20$, $p < 0.001$), and the WEF-nexus attitude ($B = 0.48$, $SE = 0.06$, $p < 0.001$) all positively predicted the increase of the WEF-nexus behavioral intention. Only family income negatively predicted it ($B = -0.28$, $SE = 0.12$, $p = 0.018$).

Effects of the Interactions on Environmental Behavioral Intention

The three-way ANOVA to analyze the contribution of the three message factors (sidedness, conclusiveness, and gain- or loss-framing) to the environmental protection behavioral intention revealed a noticeable main effect of conclusiveness ($p = 0.063$) and two arguably noticeable interactions (between conclusiveness and gain- or loss-frame, $p = 0.142$; between sidedness and conclusiveness, $p = 0.143$; Iyengar, 1991; Iyengar & Kinder, 1987). For the main effect of conclusiveness, conclusive versions generated more environmental behavioral intention ($N = 120$, $M = 48.67$, $SD = 12.13$) than the inconclusive versions ($N = 132$, $M = 45.30$, $SD = 14.31$). For the interaction between conclusiveness and gain- or loss-frame, gain-frame appeared to generate more environmental behavioral intention when conclusive ($N = 61$, $M = 47.97$, $SD = 12.37$) than inconclusive ($N = 63$, $M = 47.03$, $SD = 15.45$), and the same happened for loss-frame, just with a bigger difference (being conclusive, $N = 59$, $M = 49.39$, $SD = 11.94$; being inconclusive, $N = 69$, $M = 43.72$, $SD = 13.10$), confirming the effectiveness of conclusiveness at this level again. The loss-framed conclusive versions generated the highest mean score among the four groups, and the loss-framed inconclusive versions generated the lowest. For the interaction between sidedness and conclusiveness, two-sidedness was more effective when conclusive ($N = 58$, $M = 47.26$, $SD = 11.91$) than inconclusive ($N = 64$, $M = 46.28$, $SD = 14.51$), a pattern also happened to the one-sidedness (being conclusive, $N = 62$, $M = 49.98$, $SD = 12.28$; being inconclusive, $N = 68$, $M = 44.38$, $SD = 14.18$), again with a much bigger difference. The one-sided conclusive versions generated the highest mean score of environmental behavioral intention among the four groups, and the one-sided inconclusive versions generated the lowest. This interaction looks the same as the interaction between those two message factors captured by the three-way ANCOVA model for WEF-nexus behavioral intention.

The three-way ANCOVA controlled ideology, religiosity, WEF-nexus attitude, and WEF-nexus behavioral intention and further lowered the p-value for the main effect of conclusiveness to be statistically significant ($p = 0.019$), although no other main effects or interactions were noticeable. For the main effect of conclusiveness in increasing audience behavioral intention to protect the environment, conclusive versions ($N = 90$, $M = 49.19$, $SD = 12.85$) were clearly superior to those inconclusive ones ($N = 98$, $M = 44.98$, $SD = 14.27$). This pattern was consistent with the main effects of conclusiveness identified by the three-way ANOVA model for environmental behavioral intention and by the three-way ANCOVA model for WEF-nexus behavioral intention. It was safe to conclude that at least in this audience (American college students), on the topic of WEF nexus, conclusive versions (perhaps also with a clear call-to-action) were more persuasive than inconclusive versions in eliciting audience behavioral intention to promote the topic and protect the environment.

Three covariates, ideology (the higher the value the more liberal, $B = 3.65$, $SE = 0.90$, $p < 0.001$), WEF-nexus attitude ($B = 0.41$, $SE = 0.18$, $p = 0.027$), and WEF-nexus behavioral intention ($B = 1.09$, $SE = 0.17$, $p < 0.001$), positively contributed to the increase of environmental behavioral intention. Religiosity's contribution was not statistically significant but critical to be controlled in the model to reduce the *p*-value for the main effect of the conclusiveness.

Interpreting the Effects of the Interactions on the WEF Nexus

The three-way ANOVA models at the five levels of audience response first confirm the main effect of conclusiveness and the advantage of conclusive messages, with the main effects of sidedness and gain- or loss-framing as well as their interactions under control. The effectiveness of the conclusive messages is clearer at the behavioral levels, perhaps also because they contain a clear call-to-action element (the audience's resistance to persuasion is captured in the analyses that include the control group but not identifiable in these interactional analyses that exclude the control group). For the two-way interactions, the interaction between gain- or loss-framing and sidedness becomes statistically significant at the levels of knowledge and attitude. At the level of subjective WEF-nexus knowledge, the gain-framed messages are more effective with one-sidedness, and the loss-framed messages, two-sidedness; rendering the gain-framed one-sided messages the most persuasive among the four groups. At the levels of objective WEF-nexus knowledge and WEF-nexus attitude, however, the gain-framed messages are more effective with two-sidedness, and the loss-framed messages, one-sidedness; the loss-framed one-sided messages perform the best at the objective WEF-nexus knowledge level, and the gain-framed two-sided messages are the most persuasive at the WEF-nexus attitude level. At the two levels of behavioral intentions, the interaction between sidedness and conclusiveness is also identified: two-sided messages are more effective when inconclusive, and one-sided messages are more effective when conclusive. Strategic environmental communicators, therefore, are recommended to use one-sided conclusive messages, avoiding one-sided inconclusive messages, to promote behavioral changes related to the WEF nexus.

Two three-way interactions among all three factors are also identified by the three-way ANOVA models for WEF-nexus attitude and WEF-nexus behavioral intention, which evidences the complexity of determining persuasion approaches for environmental communications and garnering audience responses at higher levels. The two three-way interactions, however, show roughly the same pattern, indicating their robustness. For gain-framed messages, both three-way interactions show that two-sided messages are more effective when inconclusive, and one-sided messages, conclusive. For loss-framed messages, both three-way interactions also show that two-sided messages are more effective when conclusive, which can also serve as a safe rule, but one-sided messages generate more WEF-nexus attitude when inconclusive and more WEF-nexus behavioral intention when conclusive. Also, both three-way interactions reveal the gain-framed, two-sided, inconclusive message as the most persuasive among the eight message groups, and the gain-framed, one-sided, inconclusive message as the least persuasive.

The three-way ANCOVA models also confirm the main effect of conclusiveness at the levels of behavioral intentions, showing that the conclusive messages (which also contain clear call-for-action elements) are advantageous in generating behavioral intentions to support the WEF nexus and protect the environment in general. The interaction between sidedness and gain- or loss-framing is also confirmed at the levels of subjective WEF-nexus knowledge, objective WEF-nexus knowledge, WEF-nexus attitude, and probably WEF-nexus behavioral intention. Gain-framed one-sided messages, or loss-framed two-sided messages, are more effective to increase self-reported knowledge of WEF nexus than the gain-framed two-sided, or loss-framed one-sided, messages, which are effective in increasing the objective WEF-nexus knowledge and WEF-nexus behavioral intention. The most persuasive messages are the gain-framed one-sided messages at the subjective WEF-nexus knowledge level, the loss-framed one-sided messages at the objective WEF-nexus knowledge level, and the gain-framed two-sided messages at the levels of WEF-nexus attitude and WEF-nexus behavioral intention. Two-sided inconclusive messages or one-sided conclusive messages are also more effective in generating WEF-nexus attitudes than two-sided conclusive messages or one-sided inconclusive messages, with the one-sided conclusive messages being the most persuasive among these four groups.

The three-way interactions identified by the three-way ANCOVA models for objective WEF-nexus knowledge and WEF-nexus attitude are largely consistent with the major pattern of the three-way interactions identified by the three-way ANOVA models: for gain-framed messages, two-sidedness is more effective when inconclusive, and one-sidedness, conclusive; for loss-framed messages, two-sidedness is more effective when conclusive, and one-sidedness, inconclusive. It seems that, when the gain side is presented as a point of consideration, the audience prefers a message that is either deliberative (providing information from both the positive and the negative side and leaving room for the audience to think and make conclusions) or clear (providing just the positive side of the information with a clear conclusion). When the loss side is presented, however, the audience is more motivated by a message that provides some directions and also leaves room to think (offering information on both the positive and the negative sides with a clear conclusion, or offering information from just the positive side with an implicit conclusion to allow room for thinking). The most effective message to increase objective WEF-nexus knowledge is loss-framed, one-sided, and inconclusive; the most persuasive message to increase WEF-nexus attitude, however, is gain-framed, two-sided, and inconclusive, confirming that different approaches are needed to achieve strategic communication goals at different levels. The clear and consistent interactions among all three factors at the levels of objective WEF-nexus knowledge, WEF-nexus attitude, and WEF-nexus behavioral intention and between two of those factors (mainly between gain- or loss-framing and sidedness) at several levels appear robust and can guide message design for strategic environmental communicators with campaigns of different goals.

References

Ahearne, M., Gruen, T., & Saxton, M. K. (2000). When the product is complex, Does the advertisement's conclusion matter? *Journal of Business Research, 48*(1), 55–62. https://doi.org/10.1016/S0148-2963(98)00075-7

Allen, M. (1991). Meta-analysis comparing the persuasiveness of one-sided and two-sided messages. *Western Journal of Speech Communication, 55*(4), 390–404. https://doi.org/10.1080/10570319109374395

Armstrong, S. (2010). *Persuasive advertising: Evidence-based principles.* Palgrave Macmillan.

Becker, A. B., & Anderson, A. A. (2019). Using humor to engage the public on climate change: The effect of exposure to one-sided vs. two-sided satire on message discounting, elaboration and counterarguing. *JCOM, 18*(04), A07. https://doi.org/10.22323/2.18040207

Belch, G. E., & Belch, M. A. (2014). *Advertising and promotion: An integrated marketing communications perspective* (10th ed.). McGraw Hill/Irwin.

Bohner, G., Einwiller, S., Erb, H.-P., & Siebler, F. (2003). When small means comfortable: Relations between product attributes in two-sided advertising. *Journal of Consumer Psychology, 13*(4), 454–463. https://doi.org/10.1207/S15327663JCP1304_12

Carvalho, A. S. M., Godinho, C. I. A., & Graça, J. (2022). Gain framing increases support for measures promoting plant-based eating in university settings. *Food Quality and Preference, 97*, 104500. https://doi.org/10.1016/j.foodqual.2021.104500

Chen, D. (2011). *The arts of persuasion*. Enterprise Management Publishing House.

Chen, M.-F. (2016). Impact of fear appeals on pro-environmental behavior and crucial determinants. *International Journal of Advertising, 35*(1), 74–92. https://doi.org/10.1080/02650487.2015.1101908

Chen, T., Dai, M., Xia, S., & Zhou, Y. (2022). Do messages matter? Investigating the combined effects of framing, outcome uncertainty, and number format on COVID-19 vaccination attitudes and intention. *Health Communication, 37*(8), 944–951. https://doi.org/10.1080/10410236.2021.1876814

Crowley, A. E., & Hoyer, W. (1994). An integrative framework for understanding two-sided persuasion. *Journal of Consumer Research, 20*(4), 561–574. https://doi.org/10.1086/209370

de Alcantara, M., Ares, G., de Castro, I. P. L., & Deliza, R. (2020). Gain vs. loss-framing for reducing sugar consumption: Insights from a choice experiment with six product categories. *Food Research International, 136*, 109458. https://doi.org/10.1016/j.foodres.2020.109458

Dunn, S. (2013, July 8). The debates behind U.S. intervention in World War II. *The Atlantic*. Retrieved March 4, 2025, from https://www.theatlantic.com/national/archive/2013/07/the-debate-behind-us-intervention-in-world-war-ii/277572/

Eisend, M. (2006). Two-sided advertising: A meta-analysis. *International Journal of Research in Marketing, 23*, 187–198. https://doi.org/10.1016/j.ijresmar.2005.11.001

Eisend, M. (2022). The influence of humor in advertising: Explaining the effects of humor in two-sided messages. *Psychology & Marketing, 39*(5), 962–973. https://doi.org/10.1002/mar.21634

Gee, A. (2015, January 9). Who first said 'The pen is mightier than the sword'? *BBC Website*. Retrieved March 4, 2025, from https://www.bbc.com/news/magazine-30729480

Ghesla, C., Grieder, M., Schmitz, J., & Stadelmann, M. (2020). Pro-environmental incentives and loss aversion: A field experiment on electricity saving behavior. *Energy Policy, 137*, 111131. https://doi.org/10.1016/j.enpol.2019.111131

Griese, K.-M., Alexandrov, A., Michaelis, C., & Lilly, B. (2018). Examining the effect of humor in environmentally-friendly advertising. *The Marketing Management Journal, 28*(1), 30–47.

Gross, J. (2004, June 13). From guilt trip to hot wheels. *The New York Times*. Retrieved March 4, 2025, from https://www.nytimes.com/2004/06/13/style/from-guilt-trip-to-hot-wheels.html

Hameleers, M. (2021). Prospect theory in times of a pandemic: The effects of gain versus loss framing on risky choices and emotional responses during the 2020 coronavirus outbreak – Evidence from the US and the Netherlands. *Mass Communication and Society, 24*(4), 479–499. https://doi.org/10.1080/15205436.2020.1870144

Hovland, C., Lumsdaine, A. A., & Sheffield, F. D. (1949). *Experiments on mass communication*. Princeton University Press.

Hovland, C., & Mandell, W. (1952). An experimental comparison of conclusion drawing by the communicator and by the audience. *Journal of Abnormal and Social Psychology, 47*(3), 581–588. https://psycnet.apa.org/doiLanding?doi=10.1037%2Fh0059833

Iyengar, S. (1991). *Is anyone responsible: How television frames political issues*. The University of Chicago Press.

Iyengar, S., & Kinder, D. R. (1987). *News that matters: Television and American opinion*. The University of Chicago Press.

Jaccard, J., & Turrisi, R. (2003). *Interaction effects in multiple regression* (2nd ed.). Sage Publications.

Jensen, J. D., Ratcliff, C. L., Yale, R. N., Krakow, M., Scherr, C. L., & Yeo, S. K. (2018). Persuasive impact of loss and gain frames on intentions to exercise: A test of six moderators. *Communication Monographs, 85*(2), 245–262. https://doi.org/10.1080/03637751.2017.1353699

Kahneman, D. A., & Tversky, A. N. (1979). Prospect theory: An analysis of decision under risk. *Econometrica, 47*(2), 263–291. https://doi.org/10.2307/1914185

Kamins, M. A., Brand, M. J., Hoeke, S. A., & Moe, J. C. (1989). Two-sided versus one-sided celebrity endorsement: The impact on advertising effectiveness and credibility. *Journal of Advertising, 18*(2), 4–10. https://doi.org/10.1080/00913367.1989.10673146

Kardes, F. R. (1988). Spontaneous inference processes in advertising: The effects of conclusion omission and involvement on persuasion. *Journal of Consumer Research, 15*(2), 225–233. https://doi.org/10.1086/209159

Lee, Y.-K., Chang, C.-T., & Chen, P.-C. (2017). What sells better in green communications: Fear or hope? It depends on whether the issue is global or local. *Journal of Advertising Research, 57*(4), 379–396. https://doi.org/10.2501/JAR-2017-048

Levine, A. S., & Kline, R. (2019). Loss-framed arguments can stifle political activism. *Journal of Experimental Political Science, 6*(3), 171–179. https://doi.org/10.1017/XPS.2018.28

Li, S.-C., & Huang, L.-M. (2020). Fear appeals, information processing, and behavioral intentions toward climate change. *Asian Journal of Communication, 30*(3–4), 242–260. https://doi.org/10.1080/01292986.2020.1784967

Lucas, S. (2020). *The art of public speaking* (13th ed.). McGraw Hill.

Martin, B., Lang, B., & Wong, S. (2003). Conclusion explicitness in advertising: The moderating role of need for cognition (NFC) and argument quality (AQ) on persuasion. *Journal of Advertising, 32*(4), 57–65. https://doi.org/10.1080/00913367.2003.10639148

McGuire, W. J., & Papageorgis, D. (1961). The relative efficacy of various types of prior belief-defense in producing immunity against persuasion. *Journal of Abnormal Social Psychology, 62*(2), 327–337. https://doi.org/10.1037/h0042026

McKenzie-Mohr, D., Lee, N. R., Schultz, W. P., & Kotler, P. (2012). *Social marketing to protect the environment: What works.* Sage Publications.

Moyer-Guse, E., Tchernev, J. M., & Walther-Martin, W. (2019). The persuasiveness of a humorous environmental narrative combined with an explicit persuasive appeal. *Science Communication, 41*(4), 422–441. https://doi.org/10.1177/1075547019862553

Nabi, R. L., Gustafson, A., & Jensen, R. (2018). Framing climate change: Exploring the role of emotion in generating advocacy behavior. *Science Communication, 40*(4), 442–468. https://doi.org/10.1177/1075547018776019

Nan, X., Daily, K., & Qin, Y. (2018). Relative persuasiveness of gain- vs. loss-framed messages: A review of theoretical perspectives and developing an integrative framework. *Review of Communication, 18*(4), 370–390. https://doi.org/10.1080/15358593.2018.1519845

Nan, X., Madden, K., Richards, A., Holt, C., Wang, M. Q., & Tracy, K. (2016). Message framing, perceived susceptibility, and intentions to vaccinate children against HPV among African American parents. *Health Communication, 31*(7), 798–805. https://doi.org/10.1080/10410236.2015.1005280

Niederdeppe, J., Gollust, S., & Barry, C. L. (2014). Inoculation in competitive framing: Examining message effects on policy preferences. *Public Opinion Quarterly, 78*(3), 634–655. https://doi.org/10.1093/poq/nfu026

O'Keefe, D. (2016). *Persuasion: Theory and research* (3rd ed.). Sage.

O'Keefe, D. J., & Jensen, J. D. (2006). The advantages of compliance or the disadvantages of noncompliance? A meta-analytic review of the relative persuasive effectiveness of gain-framed and loss-framed messages. In C. S. Beck (Ed.), *Communication yearbook* (pp. 1–43). Lawrence Erlbaum.

O'Keefe, D. J., & Nan, X. (2012). The relative persuasiveness of gain- and loss-framed messages for promoting vaccination: A meta-analytic review. *Health Communication, 27*(8), 776–783. https://doi.org/10.1080/10410236.2011.640974

Pechmann, C. (1992). Predicting when two-sided ads will be more effective than one-sided ads: The role of correlation and correspondent inferences. *Journal of Marketing Research, 29*(4), 441–453. https://doi.org/10.2307/3172710

Redman, P., & Maples, W. (2017). *Good essay writing: A social sciences guide*. Sage Publications.

Reynolds-Tylus, T. (2019). Psychological reactance and persuasive health communication: A review of the literature. *Frontiers in Communication, 4*(56), 1–12. https://doi.org/10.3389/fcomm.2019.00056

Sawyer, A. G., & Howard, D. J. (1991). Effects of omitting conclusions in advertisements to involved and uninvolved audiences. *Journal of Marketing Research, 28*(4), 467–474. https://doi.org/10.1177/002224379102800408

Shin, S., Ki, E.-J., & Griffin, W. G. (2017). The effectiveness of fear appeals in 'green' advertising: An analysis of creative, consumer, and source variables. *Journal of Marketing Communications, 23*(5), 473–492. https://doi.org/10.1080/13527266.2017.1290671

Szabo, E. A., & Pfau, M. (2002). Nuance in inoculation: Theory and applications. In J. Dillard & M. Pfau (Eds.), *The persuasion handbook: Development in theory and practice* (pp. 233–258). Sage.

Velde, F. V., Hudders, L., Cauberghe, V., & Claerebout, E. (2017). Changing farmers' behavior intention with a hint of wit: The moderating influence of humor on message sidedness. *Journal of Environmental Psychology, 56*, 97–103. https://doi.org/10.1016/j.jenvp.2017.12.001

Yao, Q. (2007). China's official framing of religion and its influence on young Chinese students: A partial testing of the process model of framing in a special media environment. *Asian Journal of Communication, 17*(4), 416–432. https://doi.org/10.1080/01292980701637033

Yao, Q. (2021). Informing, implying or directing? Testing the effects of message sidedness, conclusiveness and their interaction in national and local Google Ads campaigns. *Journal of Research in Interactive Marketing, 15*(4), 623–640. https://doi.org/10.1108/JRIM-09-2019-0141

Yao, Q., Haggard, C., & Cai, L. (2014). Advertising sidedness and conclusiveness across cultures: How Chinese and U.S. college students are influenced by different climate-change ads. *Chinese Media Research, 10*(2), 86–94. https://www.chinamediaresearch.net/readmore/vol10no2/CMR140209Yao.jpg

Yao, Q., Joshi, P., Chang, C.-F., McDonalds, C., Tran, J., Wheeler, W., & Hou, S. (2018). Advocating a new approach to governing water, energy, and food security: Testing the effects of message inoculation and conclusion explicitness in the case of the WEF nexus. *Journal of International Crisis and Risk Communication Research, 1*(1), 111–133. https://jicrcr.com/index.php/jicrcr/article/view/6/6

CHAPTER 7

Conclusions and the Next Step to Promote the WEF Nexus

Strategic environmental communication has drastically emerged as an urgent research topic (Cox & Depoe, 2015; Pezzullo & Cox, 2021; Yao et al., 2020) due to the devastating environmental issues such as climate change (Yao et al., 2014; Yao et al., 2024) and water, energy, and food shortages (BBC, 2022, May 25; Hoff, 2011; Kaiman, 2014; Kann et al., 2021; United Nations, 2014; UN News, 2022, August 2; World Economic Forum, 2011). The need is pressing for effective campaigns to raise public awareness and support for globally solving environmental issues (Brulle, 2010; Yao et al., 2020), which are critical to allocating financial and moral resources to research, planning, and execution of solutions to those issues (Bhaduri et al., 2015; Obermiller, 1995; Scott et al., 2015; Yao et al., 2018). This book, based on the National College Student Science Literacy Survey (NCSSLS) data, has tested the functions of several message factors and levels to promote the water-energy-food (WEF) nexus, an emerging concept of governing water, energy, and food systematically, to college students in the U.S. Its findings can also help design strategic communication messages for other new environmental topics and understand how audiences process persuasion in general.

Q. J. Yao, *Marketing the Water-Energy-Food (WEF) Nexus*,
https://doi.org/10.1007/978-3-031-93148-2_7

1 Conclusions on the Levels of Message Effects and the WEF Nexus

The random assignment of 402 subjects with valid responses from 185 universities/colleges across the U.S. into 11 groups exposed to a version of the WEF-nexus message with components at different levels (agenda, knowledge, attitudinal, or behavioral) and a control group in the second NCSSLS round enables testing the five-level model of message effects (Yao, 2008; Yao et al., 2020) developed from the research of the hierarchy of audience responses (Lavidge & Steiner, 1961; Leckenby & Wedding, 1982; McGuire, 1989, 2013; Rogers, 2003; Xu et al., 2010) and mass media effects (e.g., Cacciatore et al., 2016; McCombs & Reynolds, 2009; Moy et al., 2016; Roskos-Ewoldsen et al., 2002).

The positive associations between the levels in the hierarchy of audience responses (agenda, knowledge/frame, attitude, and behavioral intention) are largely confirmed. The positions of the concepts of audience WEF-nexus agenda and frame in the hierarchy, however, need further studies because they are associated with the audience's WEF-nexus attitude and behavioral intention, not objective WEF-nexus knowledge. Furthermore, the WEF-nexus agenda is theoretically ranked lower than the WEF-nexus frame in the hierarchy of audience responses but is, in this dataset, more strongly associated with the audience's WEF-nexus attitude and behavioral intention than the audience's WEF-nexus frame. So, the hierarchy of audience responses seems the right way to understand audience information processing and responses, but a continual explication of the concepts in the hierarchy is still needed to obtain more accurate, standardized, and measurable concepts.

In the five-level model of message effects (Yao, 2008; Yao et al., 2020), only the knowledge level of message effects is confirmed in these data, indicating that message effects start to build from the lower levels. The agenda-setting effect is not identified possibly because of, again, the traditional explication of the audience's WEF-nexus agenda. In addition, the lack of evidence supporting other levels of the media effects in the model can also be due to the small magnitude of the single-message stimulus in the study's noisy natural setting, which was chosen for more external generalizability. In the longer term, however, the audience's "keeping up with science news" unequivocally positively contributes to the audience's WEF-nexus agenda, objective knowledge, attitude, and behavioral intention, substantially confirming the five-level model of message effects. Meanwhile, the five-level model of message effects (Yao, 2008; Yao et al., 2020) is synthesized from decades of multi-field studies of the hierarchy of audience responses and media effects. While the model needs further testing to improve, it can still be instrumental to help strategic environmental communicators understand the audience's processing of, and responding to, messages on topics such as the WEF nexus.

2 Conclusions on Effects of Message Factors and the WEF Nexus

The survey experiment embedded in the NCSSLS second round was also designed to test the persuasiveness of three message factors: sidedness (Crowley & Hoyer, 1994; Eisend, 2006; Hovland et al., 1949; Kamins et al., 1989; O'Keefe, 2016; Pechmann, 1992), conclusiveness (Armstrong, 2010; Hovland & Mandell, 1952; O'Keefe, 2016), and gain- or loss-frame (Hameleers, 2021; Kahneman & Tversky, 1979; O'Keefe & Jensen, 2006; O'Keefe & Nan, 2012).

2.1 Message Sidedness and the WEF Nexus

Are messages more persuasive when focusing solely on supporting arguments, or when presenting both supporting and opposing viewpoints? Literature on inoculation theory upholds that two-sided messages refuting the opposing points are more persuasive than two-sided messages without refutation or one-sided messages (McGuire & Papageorgis, 1961; Niederdeppe et al., 2014; Szabo & Pfau, 2002). Studies in this book, however, suggest that, for the relatively new concept of WEF nexus, one-sided messages are advantageous overall, but two-sided messages with refutation can be more effective at certain levels of message effects when controlling some confounders.

Specifically, when just comparing the one-sided messages, the two-sided messages, and the control group, the one-sided messages seem to increase the audience's subjective knowledge of the WEF nexus above the level of the control group, but the two-sided messages have no such effect. Both the one-sided messages and the two-sided messages can increase the audience's objective knowledge above the level of the control group, but one-sided messages seemingly increase it even higher. No differences are observed at the levels of WEF-nexus attitude and behavioral intention as well as the environmental behavioral intention. Again, communication effects begin at the knowledge level, and one-sided messages are advantageous at this level for conveying clearer information.

When confounding demographic and socioeconomic characteristics of the audience are controlled, it is even clearer that both versions of the message can increase the audience's subjective or objective WEF-nexus knowledge, with one-sided messages showing more advantages. The audience's psychological reactance to persuasion is detected at the levels of WEF-nexus attitude and behavioral intention in this study, which lowers the mean scores of the message groups compared with that of the control group. Facing such resistance, one-sided messages are undoubtedly a better choice than their two-sided counterparts, probably due to the absence of negativity. One-sided messages, nevertheless, can lower the audience's behavioral intention toward the WEF nexus even more, possibly because they somehow synergize with the audience's psychological reactance to persuasion.

The implication of these results is the need to establish and analyze the goals of strategic environmental communications for new concepts, such as the WEF nexus, before designing message strategies. Cognitive knowledge should be provided at the initial part of strategic environmental campaigns, which prefers the one-sided strategy. The two-sided strategy should be used when resistance to persuasion is identified and the goal of communication is to generate audience behaviors (Crowley & Hoyer, 1994; Eisend, 2006; Kamins et al., 1989; O'Keefe, 2016; Pechmann, 1992).

2.2 *Message Conclusiveness and the WEF Nexus*

Strategic environmental communicators also wonder whether they should "provide a conclusion" when designing persuasive messages (Armstrong, 2010; O'Keefe, 2016). The tests conducted in this book reveal that, overall, inconclusive messages increase the subjective knowledge of the WEF nexus more, and conclusive messages increase the objective knowledge more. With covariates controlled, both versions of messages certainly drive both the subjective and the objective knowledge above the level of the control group.

No differences among the conclusive message, inconclusive message, and control groups can be detected at the levels of WEF-nexus attitude and behavioral intention. When confounders are under control, however, resistance to persuasion appears again, lowering the mean scores for the message groups compared to those of the control group. With audience resistance, the conclusive messages appear to be more persuasive than the inconclusive messages at the levels of WEF-nexus attitude and behavioral intention and environmental behavioral intention.

Those results generally support the conclusion of previous research that conclusive messages are more persuasive than their inconclusive counterparts (O'Keefe, 2016), particularly on issues not hotly debated. Strategic environmental communicators should choose the conclusive strategy when designing messages to promote issues such as the WEF nexus.

2.3 *Gain- or Loss-Framing and the WEF Nexus*

Gain- or loss-framing research, derived from the prospect theory (Kahneman & Tversky, 1979, 1984), advises strategic environmental communicators to create loss-framed messages to address the audience's loss-aversion psychology. This advice, however, is rejected in this study when the gain-framed messages show more effectiveness in increasing the audience's self-perceived knowledge of the WEF nexus. At the objective WEF-nexus knowledge level, both the gain- and the loss-framed messages seem to increase the mean scores over the control group. No differences can be detected at the WEF-nexus attitude or behavioral intention levels. Hence, overall, the prospect theory is not supported by this dataset.

The results change when confounders are controlled in the ANCOVA models. Although both the gain- and loss-framed messages again increase both the subjective and the objective WEF-nexus knowledge above the control group with little difference shown between themselves, the loss-framed messages show signs of being more persuasive than the gain-framed messages at the level of WEF-nexus attitude. At the level of WEF-nexus behavioral intention, suffering from the audience's reactance to persuasion, the gain-framed messages obviously lower their mean scores below the control group, while the loss-framed messages do not, indicating an advantage. So, the prospect theory finds support at the levels of WEF-nexus attitude and behavioral intention when controlling confounders. At the environmental behavioral intention level, again with the audience's possible resistance to persuasion, however, the gain-frame seems more persuasive than the loss-frame.

Those results confirm the complexity of message effects, as different message factors function differently at different levels of audience responses. With these data and confounders controlled, the prospect theory seems supported at the levels of WEF-nexus attitude and behavioral intention but reversed essentially at the subjective knowledge level and perhaps the environmental behavioral intention level. While the general pattern is noticeable for strategic communicators, more research is needed to understand how confounders affect the persuasiveness of the gain- or loss-frame at different levels of audience responses (Jensen et al., 2018; Nan et al., 2018; Reynolds-Tylus, 2019).

2.4 *The Interactions of Message Factors and the WEF Nexus*

Likely the first study to examine message sidedness, conclusiveness, and gain- or loss-frame together, the NCSSLS data provide insights into their interactions with each other at different levels of audience responses.

The advantageous effects of conclusive messages are the first finding of the interaction among these three message factors. When message sidedness and gain- or loss-frame are controlled, sometimes together with other covariates, the main effects of conclusiveness become evident at the levels of behavioral intentions. Conclusive messages are apparently more persuasive in generating higher levels of WEF-nexus behavioral intention and environmental behavioral intention than inconclusive messages, when the control group is excluded from the analysis to identify the probable audience reactance.

Sidedness and gain- or loss-frame interact at the levels of subjective WEF-nexus knowledge, objective WEF-nexus knowledge, WEF-nexus attitude, and probably WEF-nexus behavioral intention, with different interactional patterns. At the level of subjective WEF-nexus knowledge, gain-frame is more effective when one-sided, and loss-frame is more effective when two-sided; one-sided gain-framed messages are the most persuasive, and one-sided loss-framed messages are the least. With objective WEF-nexus knowledge, WEF-nexus attitude, and WEF-nexus behavioral intention, the gain-frame is more persuasive when two-sided, and the loss-frame is more persuasive when one-sided. The most and the least persuasive messages, however, are different: One-sided loss-framed messages are the most persuasive, and two-sided loss-framed messages the least, for objective WEF-nexus knowledge; for both WEF-nexus attitude and behavioral intention, the most persuasive is the two-sided gain-frame, but the least persuasive for the WEF-nexus attitude is the two-sided loss-frame and that for the WEF-nexus behavioral intention is the one-sided gain-frame.

Sidedness and conclusiveness interact at the levels of WEF-nexus behavioral intention and, probably, WEF-nexus attitude and environmental behavioral intention. At the levels of WEF-nexus attitude and behavioral intention, two-sided messages are more persuasive when inconclusive, and one-sided messages are more persuasive when conclusive. At the level of environmental behavioral intention, both the two-sided and the one-sided messages are more persuasive when conclusive. While at all three levels, the one-sided inconclusive messages are the least persuasive, the most persuasive at the level of WEF-nexus attitude are the two-sided inconclusive messages, and at the two levels of behavioral intentions are the one-sided conclusive messages.

Three-way interactions are identified at the levels of WEF-nexus attitude and behavioral intention and, probably also, objective WEF-nexus knowledge. Their patterns are largely the same, but the best and worst performers again vary: for WEF-nexus attitude and behavioral intention, the gain-framed, two-sided, inconclusive message is the most persuasive among the eight versions, and the gain-framed, one-sided, inconclusive message is the least; for objective WEF-nexus knowledge, the loss-framed, one-sided, inconclusive message seems the most persuasive, and the loss-framed, two-sided, inconclusive message seems the least. Strategic environmental communicators need to pay attention to these interactions when combining these factors in their messages.

3 The Next Step: Interactions, Levels of Responses, and Causality

The drastically deteriorating living conditions on Earth demand effective strategic environmental communications to raise public awareness and support for sustainable living styles and policies globally (BBC, 2022, May 25; Scott et al., 2015; Stillwell, 2015; UN News, 2022, August 2). Effective communication depends on a thorough understanding of message effects and audience responses. Despite a significant volume of literature accumulation, key questions remain unaddressed, and future research is urgently needed at a time when global policy attention to environmental issues has dropped.

For future research on message effects and audience responses, particularly when related to new environmental issues such as the WEF nexus, the present study provides three major suggestions.

First, variables of message factors and audience responses need to be explicated clearly and standardized so that message effects can be analyzed at different levels (McGuire, 1989; Yao et al., 2020). Confusion and the mixed results of existing studies are often generated by measuring message effects at different levels with different methods but labeling them as the same variable (Yao, 2024). The five-level model of message effects, synthesized from decades of extant studies of media effects and audience responses (Yao et al., 2020), provides fundamental guidance for future research on message effects and practices in strategic communications. Its levels of audience agenda and frame, however, need clearer conceptualization and operationalization.

Second, the interactions of message factors need further examination. A body of research has been done on the persuasiveness of single message factors or their interaction with audience characteristics (Chen et al., 2022; Crowley & Hoyer, 1994; Eisend, 2006; Jensen et al., 2018; Kamins et al., 1989; Martin et al., 2003; Nabi et al., 2018; Nan et al., 2016; O'Keefe, 2016; Pechmann, 1992). The NCSSLS data, together with other studies (Yao, 2021; Yao et al., 2014; Yao et al., 2018), reveal that message factors, often combined in messages, can interact to alter the messages' persuasiveness, which needs to be sufficiently understood for effective copywriting.

Third, the concept of causation needs revisiting. Researchers and practitioners of strategic communications, just like most scientists, are committed to identifying causalities (Hitchcock, 2008; Hunt, 2015; O'Shaughnessy, 1992), such as the relationship between message stimuli and audience responses. Causation demands a substantial (nonspurious) temporal concomitant variation (covariation; Liu et al., 2021; Whisman et al., 2021; Wunsch et al., 2021) that necessitates a statistical correlation. The analyses in this book show, however, that when different variables are controlled, the patterns of correlation can change dramatically. Including or excluding just one variable can occasionally substantially change the correlations or regression coefficients of other variables in the model. These results posit a thought on causality. On one side, the traditional thinking of the cause as an independent variable changing the dependent variable, as implied in the casual theories of regularity, probability, counterfactuality (but for, sine qua non), manipulability/intervention, or process (Hitchcock, 2008; Pearl & Mackenzie, 2018), is merely a myth or "a relic of a bygone age" (Hitchcock, 2008, p. 317; Wittgenstein, 1922/2025) in scientific research. No variables can single-handedly change a response variable so robustly that it can be identified as the cause as we call it. As more variables are excluded from the analysis, fewer variables remain correlated with the response variable; just like, keeping removing environmental factors such as air or air pressure, fire alone cannot boil water in a bowl. On the other hand, with the increasing realization of the overuse and limitation of experimentation, the "gold-standard method of causation-identification" (Nagin & Sampson, 2019), and the prevalence of big data, substantial correlations revealed in studies are often sufficient to explain and predict phenomena. Identifying causality is both impossible and unnecessary in scientific research, which instead aims to identify essential correlations within a context, system, synergy (defined by the covariates in the analyses), or a causal mechanism (Hedstrom & Ylikoski, 2010) for explanation or prediction. Message factors or levels alone may not change audience responses, but they may with certain covariates, whose contribution sometimes is even not statistically significant. Strategic communicators need to pay attention to those contexts when understanding the persuasiveness of message factors or levels of stimuli.

The WEF nexus, as a newer approach to governing the natural resources of water, energy, and food systematically and coordinately, has been deemed by several leading international organizations and many scientists as effective in solving the shortages and crises of those resources globally (e.g., Hoff, 2011; United Nations, 2014; World Economic Forum, 2011). However, the concept is still largely unknown to the public (Yao et al., 2018; Yao et al., 2024), including college students, an educated group. Effective messages need to be designed to communicate the concept to the public to avoid politicization and garner support to facilitate research, planning, and execution of the nexus solutions. This book, likely the first one to examine effective communication of the WEF nexus, serves as the foundation as well as the initial call for more research on the WEF nexus and its effective communication to help the globe solve the resource crises that threaten our very lives on this Earth.

References

Armstrong, S. (2010). *Persuasive advertising: Evidence-based principles.* Palgrave Macmillan.

BBC. (2022, May 25). NeWater: Turning urine into beer in Singapore. *BBC Website.* Retrieved May 26, 2022, from https://www.bbc.com/news/av/world-asia-61577976

Bhaduri, A., Ringler, C., Dombrowski, I., Mohtar, R., & Scheumann, W. (2015). Sustainability in the water-energy-food nexus. *Water International, 40*(5–6), 723–732. https://doi.org/10.1080/02508060.2015.1096110

Brulle, R. J. (2010). From environmental campaigns to advancing the public dialog: Environmental communication for civic engagement. *Environmental Communication, 4*(1), 82–98. https://doi.org/10.1080/17524030903522397

Cacciatore, M., Scheufele, D., & Iyengar, S. (2016). The end of framing as we know it… and the future of media effects. *Mass Communication and Society, 19*(1), 7–23. https://doi.org/10.1080/15205436.2015.1068811

Chen, T., Dai, M., Xia, S., & Zhou, Y. (2022). Do messages matter? Investigating the combined effects of framing, outcome uncertainty, and number format on COVID-19 vaccination attitudes and intention. *Health Communication, 37*(8), 944–951. https://doi.org/10.1080/10410236.2021.1876814

Cox, R., & Depoe, S. (2015). Emergence and growth of the 'field' of environmental communication. In A. Hansen & R. Cox (Eds.), *The Routledge handbook of environment and communication (Chapter 1)*. Routledge. https://doi.org/10.4324/9781315887586

Crowley, A. E., & Hoyer, W. (1994). An integrative framework for understanding two-sided persuasion. *Journal of Consumer Research, 20*(4), 561–574. https://doi.org/10.1086/209370

Eisend, M. (2006). Two-sided advertising: A meta-analysis. *International Journal of Research in Marketing, 23*, 187–198. https://doi.org/10.1016/j.ijresmar.2005.11.001

Hameleers, M. (2021). Prospect theory in times of a pandemic: The effects of gain versus loss framing on risky choices and emotional responses during the 2020 coronavirus outbreak – Evidence from the US and the Netherlands. *Mass Communication and Society, 24*(4), 479–499. https://doi.org/10.1080/15205436.2020.1870144

Hedstrom, P., & Ylikoski, P. (2010). Causal mechanisms in social sciences. *Annual Review of Sociology, 36*, 49–67. https://doi.org/10.1146/annurev.soc.012809.102632

Hitchcock, C. (2008). Causation. In S. Psillos & M. Curd (Eds.), *The Routledge companion to philosophy of science* (pp. 317–326). Routledge.

Hoff, H. (2011). Understanding the nexus: Background paper for the Bonn 2011 Nexus Conference. SEI Website, November 16–18, 2011. https://uploads.water-energy-food.org/resources/SEI-Paper-Hoff-UnderstandingTheNexus-2011.pdf

Hovland, C., Lumsdaine, A. A., & Sheffield, F. D. (1949). *Experiments on mass communication*. Princeton University Press.

Hovland, C., & Mandell, W. (1952). An experimental comparison of conclusion drawing by the communicator and by the audience. *Journal of Abnormal and Social Psychology, 47*(3), 581–588. https://psycnet.apa.org/doiLanding?doi=10.1037%2Fh0059833

Hunt, S. D. (2015). *Marketing theory: Foundations, controversy, strategy, resource-advantage theory*. Routledge.

Jensen, J. D., Ratcliff, C. L., Yale, R. N., Krakow, M., Scherr, C. L., & Yeo, S. K. (2018). Persuasive impact of loss and gain frames on intentions to exercise: A test of six moderators. *Communication Monographs, 85*(2), 245–262. https://doi.org/10.1080/03637751.2017.1353699

Kahneman, D., & Tversky, A. (1984). Choice, values, and frames. *American Psychologist, 39*(4), 341–350. https://doi.org/10.1037/0003-066X.39.4.341

Kahneman, D. A., & Tversky, A. N. (1979). Prospect theory: An analysis of decision under risk. *Econometrica, 47*(2), 263–291. https://doi.org/10.2307/1914185

Kaiman, J. (2014, December 12). China's water diversion project starts to flow to Beijing. *The Guardian*. Retrieved March 4, 2025, from https://www.theguardian.com/world/2014/dec/12/china-water-diversion-project-beijing-displaced-farmers

Kamins, M. A., Brand, M. J., Hoeke, S. A., & Moe, J. C. (1989). Two-sided versus one-sided celebrity endorsement: The impact on advertising effectiveness and credibility. *Journal of Advertising, 18*(2), 4–10. https://doi.org/10.1080/00913367.1989.10673146

Kann, D., Rigdon, R., & Wolfe, D. (2021, August 21). The Southwest's most important river is drying up. *CNN Website*. Retrieved March 4, 2025, from https://www.cnn.com/interactive/2021/08/us/colorado-river-water-shortage/

Lavidge, R. J., & Steiner, G. A. (1961). A model for predictive measurements of advertising effectiveness. *Journal of Marketing, 25*(6), 59–62. https://doi.org/10.1177/002224296102500611

Leckenby, J., & Wedding, N. (1982). *Advertising management: Criteria, analysis, and decision making*. Grid Publishing.

Liu, T., Ungar, L., & Kording, K. (2021). Quantifying causality in data science with quasi-experiments. *Nature Computational Science, 1*, 24–32. https://doi.org/10.1038/s43588-020-00005-8

Martin, B., Lang, B., & Wong, S. (2003). Conclusion explicitness in advertising: The moderating role of need for cognition (NFC) and argument quality (AQ) on persuasion. *Journal of Advertising, 32*(4), 57–65. https://doi.org/10.1080/00913367.2003.10639148

McCombs, M., & Reynolds, A. (2009). News influence on our pictures of the world. In J. Bryant & D. Zillmann (Eds.), *Media effects: Advances in theory and research* (3rd ed., pp. 1–16). Routledge.

McGuire, W. J. (1989). Theoretical foundations of campaigns. In R. E. Rice & C. K. Atkin (Eds.), *Public communication campaigns* (2nd ed., pp. 43–65). Sage.

McGuire, W. J. (2013). McGuire's classic input-output framework for constructing persuasive messages. In R. E. Rice & C. K. Atkin (Eds.), *Public communication campaigns* (4th ed., pp. 133–146). Sage.

McGuire, W. J., & Papageorgis, D. (1961). The relative efficacy of various types of prior belief-defense in producing immunity against persuasion. *Journal of Abnormal Social Psychology, 62*(2), 327–337. https://doi.org/10.1037/h0042026

Moy, P., Tewksbury, D., & Rinke, E. M. (2016). Agenda-setting, priming, and framing. In K. B. Jensen & R. T. Craig (Eds.), *The international encyclopedia of communication theory and philosophy* (Vol. 1, pp. 52–64). Wiley. https://doi.org/10.1002/9781118766804.wbiect266

Nabi, R. L., Gustafson, A., & Jensen, R. (2018). Framing climate change: Exploring the role of emotion in generating advocacy behavior. *Science Communication, 40*(4), 442–468. https://doi.org/10.1177/1075547018776019

Nagin, D. S., & Sampson, R. J. (2019). The real gold standard: Measuring counterfactual worlds that matter most to social science and policy. *Annual Review of Criminology, 2*, 123–145. https://doi.org/10.1146/annurev-criminol-011518-024838

Nan, X., Daily, K., & Qin, Y. (2018). Relative persuasiveness of gain- vs. loss-framed messages: A review of theoretical perspectives and developing an integrative framework. *Review of Communication, 18*(4), 370–390. https://doi.org/10.1080/15358593.2018.1519845

Nan, X., Madden, K., Richards, A., Holt, C., Wang, M. Q., & Tracy, K. (2016). Message framing, perceived susceptibility, and intentions to vaccinate children against HPV among African American parents. *Health Communication, 31*(7), 798–805. https://doi.org/10.1080/10410236.2015.1005280

Niederdeppe, J., Gollust, S., & Barry, C. L. (2014). Inoculation in competitive framing: Examining message effects on policy preferences. *Public Opinion Quarterly, 78*(3), 634–655. https://doi.org/10.1093/poq/nfu026

O'Keefe, D. (2016). *Persuasion: Theory and research* (3rd ed.). Sage.

O'Keefe, D. J., & Jensen, J. D. (2006). The advantages of compliance or the disadvantages of noncompliance? A meta-analytic review of the relative persuasive effectiveness of gain-framed and loss-framed messages. In C. S. Beck (Ed.), *Communication yearbook* (pp. 1–43). Lawrence Erlbaum.

O'Keefe, D. J., & Nan, X. (2012). The relative persuasiveness of gain- and loss-framed messages for promoting vaccination: A meta-analytic review. *Health Communication, 27*(8), 776–783. https://doi.org/10.1080/10410236.2011.640974

O'Shaughnessy, J. (1992). *Explaining buyer behavior: Central concepts and philosophy of science issues*. Oxford University Press.

Obermiller, C. (1995). The baby is sick/the baby is well: A test of environmental communication appeals. *Journal of Advertising, 24*(2), 55–70. https://doi.org/10.1080/00913367.1995.10673476

Pearl, J., & Mackenzie, D. (2018). *The book of why: The new science of cause and effect*. Basic Books.

Pechmann, C. (1992). Predicting when two-sided ads will be more effective than one-sided ads: The role of correlation and correspondent inferences. *Journal of Marketing Research, 29*(4), 441–453. https://doi.org/10.2307/3172710

Pezzullo, P., & Cox, R. (2021). *Environmental communication and the public sphere* (6th ed.). Sage.

Reynolds-Tylus, T. (2019). Psychological reactance and persuasive health communication: A review of the literature. *Frontiers in Communication, 4*(56), 1–12. https://doi.org/10.3389/fcomm.2019.00056

Rogers, E. (2003). *Diffusion of innovation* (5th ed.). Free Press.

Roskos-Ewoldsen, D. R., Roskos-Ewoldsen, B., & Carpentier, F. (2002). Media priming: A synthesis. In J. Bryant & D. Zillmann (Eds.), *Media effects: Advances in theory and research* (2nd ed., pp. 97–120). Lawrence Erlbaum.

Scott, C., Kurian, M., & Wescoat, J. L., Jr. (2015). The water-energy-food nexus: Enhancing adaptive capacity to complex global challenges. In M. Kurian & R. Ardakanian (Eds.), *Governing the nexus: Water, soil, and waste resources considering global change* (pp. 15–38). Springer International Publishing. https://doi.org/10.1007/978-3-319-05747-7_2

Stillwell, A. (2015). Sustainability of public policy: Example from the energy-water nexus. *Journal of Water Resources Planning and Management, 141*(12), A4015001. https://doi.org/10.1061/(ASCE)WR.1943-5452.0000522

Szabo, E. A., & Pfau, M. (2002). Nuance in inoculation: Theory and applications. In J. Dillard & M. Pfau (Eds.), *The persuasion handbook: Development in theory and practice* (pp. 233–258). Sage.

UN News. (2022, August 2). American West faces water and power shortages due to climate crisis: UN environment agency. *United Nations Website*. Retrieved March 4, 2025, from https://news.un.org/en/story/2022/08/1123782

United Nations. (2014). The water-energy-food nexus: A new approach in support of food security and sustainable agriculture. *FAO.org*. Retrieved March 4, 2025, from http://www.fao.org/3/a-bl496e.pdf

Whisman, M. A., Sbarra, D. A., & Steven, R. H. (2021). Intimate relationships and depression: Searching for causation in the sea of associations. *Annual Review of Clinical Psychology, 17*, 233–258. https://doi.org/10.1146/annurev-clinpsy-081219-103323

Wittgenstein, L. (1922/2025). *Tractatus logico-philosophicus* (Side-by-Side ed.) Trans. Ogden/Ramsey; Pears/McGuinness. Kegan Paul. http://people.umass.edu/klement/tlp/

World Economic Forum. (2011). *Water security: The water–energy–food–climate nexus*. Island Press. http://www3.weforum.org/docs/WEF_WI_WaterSecurity_WaterFoodEnergyClimateNexus_2011.pdf

Wunsch, G., Russo, F., Mouchart, M., & Orsi, R. (2021). Time and causality in social sciences. *Time & Society, 31*(2), 157–176. https://doi.org/10.1177/0961463X211029488

Xu, W., Sun, G., Lin, Z., Chen, M., Yang, B., Chen, H., & Cao, K. (2010). Knowledge, attitude, and behavior in patients with atrial fibrillation undergoing radiofrequency catheter ablation. *Journal of Interventional Cardiac Electrophysiology, 28*(3), 199–207. https://doi.org/10.1007/s10840-010-9496-2

Yao, Q. (2008). Exploring the social dynamics in the U.S. democracy: Presidential and public opinions about, and media coverage of, environmental issues. Doctoral dissertation, University of South Carolina.

Yao, Q. (2021). Informing, implying or directing? Testing the effects of message sidedness, conclusiveness and their interaction in national and local Google Ads campaigns. *Journal of Research in Interactive Marketing, 15*(4), 623–640. https://doi.org/10.1108/JRIM-09-2019-0141

Yao, Q. (2024). Concepts and reasoning: A conceptual review and analysis of logical issues in empirical social science research. *Integrative Psychological and Behavioral Science, 58*(2), 502–530. https://doi.org/10.1007/s12124-023-09792-x

Yao, Q., Chang, C.-F., Joshi, P., & McDonald, C. (2024). Climate change versus the water-energy-food nexus: The oldness or newness of the scientific issue as a factor in the deficit model and the hierarchy of response model. *Environment, Development and Sustainability, 26*, 823–840. https://doi.org/10.1007/s10668-022-02735-3

Yao, Q., Haggard, C., & Cai, L. (2014). Advertising sidedness and conclusiveness across cultures: How Chinese and U.S. college students are influenced by different climate-change ads. *Chinese Media Research, 10*(2), 86–94. https://www.chinamediaresearch.net/readmore/vol10no2/CMR140209Yao.jpg

Yao, Q., Joshi, P., Chang, C.-F., McDonalds, C., Tran, J., Wheeler, W., & Hou, S. (2018). Advocating a new approach to governing water, energy, and food security: Testing the effects of message inoculation and conclusion explicitness in the case of the WEF nexus. *Journal of International Crisis and Risk Communication Research, 1*(1), 111–133. https://jicrcr.com/index.php/jicrcr/article/view/6/6

Yao, Q., Liu, Z., & Stephens, L. F. (2020). Exploring the dynamics in the environmental discourse: The longitudinal interaction among public opinion, presidential opinion, media coverage, policymaking in 3 decades and an integrated model of media effects. *Environmental Systems and Decisions, 40*(1), 14–28. https://doi.org/10.1007/s10669-019-09746-y

Appendix: Additional Information on the National College Student Science Literacy Survey

First Notice for the Survey

Dear [FirstName—information in these square brackets was dynamically inserted by the survey software] [LastName]:

You are randomly selected to participate in the National Science Survey of College Students conducted by us at xxx University in xxx. According to the U.S. News & World Report, [university/college population N] students have enrolled in [university/college name]. We therefore randomly selected [university/college sample N] students from [university/college name] into our sample, and you are one of them representing your college in this survey.

Your opinion is critical to us and people who want to understand college students' opinions on some scientific issues. We appreciate your participation, which is totally voluntary and can be withdrawn at any time with no penalty. No more than minimal risks are anticipated in your participation.

The panel survey contains two rounds of surveys, the first round in the week of April 3–7, 2017, and the follow-up round in the week of May 1–5, 2017, which use mostly the same questions. Each round may be emailed to you and take you about 10–15 minutes to complete. Five newsletters will be sent to you between those two rounds of surveys to keep you updated on the topic of the survey.

Q. J. Yao, *Marketing the Water-Energy-Food (WEF) Nexus*,
https://doi.org/10.1007/978-3-031-93148-2

If you participate in the panel surveys, you will have a chance to receive an iPad Air 2 valued at about $399. For the first round, 11 iPhones are given to the 1st, 10th, 360th, 700th, 1000th, 1500th, 2000th, 2500th, 3000th, 3500th, and 4000th takers of the round. If your ID generated by the online survey software matches one of the above 11 survey-respondent IDs, you will be contacted through email for receiving an iPad Air 2 that will be mailed to you by April 14, 2017.

The follow-up round of the survey will be administered in March 2017. Eleven iPad Air 2s will also be given to the 1st, 15th, 350th, 650th, 1000th, 1500th, 2000th, 2500th, 3000th, 3500th, and 4000th takers in that round. The 11 iPad Air 2s for the follow-up round will be mailed to recipients by May 13, 2017. Receiving an iPad Air 2 from the first round does not prohibit you from receiving an iPad Air 2 from the follow-up round of the survey.

In both rounds of surveys, our sample has 4013 participants randomly selected nationwide, and your chance of receiving an iPad Air 2 is 1 to about 365. Receivers may need to provide necessary information to xxx University's finance department before the iPad Air 2s are shipped.

Your email address is collected from the directory of your university or provided by your university and is only used to send you the survey link, survey-related materials, and notice if you will receive an iPad Air 2.

All information collected will be only used for our research, kept confidential, and analyzed at an aggregated level. In order to maintain data confidentiality, only the lead principal investigator has access to the survey website. Once the data are downloaded, the emails will be immediately removed from the dataset for only faculty researchers to analyze. No individual information will be released in any case. We will share the results of this study with you if you request.

The study has been approved by the xxx University IRB. Please click the link below to the IRB Approval letter [link omitted].

Message Versions Embedded in the Second Round

1. The Agenda Version (Title Only with No Message Body, One-Sided, Closed-Ended, Gain-Framed)

The number of media messages about the water-energy-food nexus, an approach to achieve water, energy, and food security simultaneously, has increased quickly since the approach was proposed in 2011.

2. Knowledge Version (One-Sided, Closed-Ended, Gain-Framed)

Water, Energy, and Food Security Is in a Nexus to Achieve!

Water, energy, and food are essentially linked. Water is needed to produce the crops and cook the foods. Agriculture is currently the largest user of water, consuming 70% of global water withdrawal. Energy is required to produce and distribute water and food. Just the food production and supply chain accounts for about 30% of total global energy consumption. Finally, the generation of energy consumes a large share of water withdrawal.

The linkage among water, energy, and food creates interactions and trade-offs. Watering crops to promote food production reduces river flows and hydropower potential. Growing bioenergy crops, such as corn for fuel ethanol, demands more water withdrawal and hurts food security. To make the watering system more efficient, such as using drip, center pivot, or rain guns, saves water but consumes more energy. The global community is well aware of water, energy, and food challenges but has so far addressed them separately, with fragmented responsibilities, lack of coordination, and inconsistencies among laws and regulations.

However, balancing those benefits is central to jointly ensuring water, energy, and food security at the same time. To do so, decision-makers, even those responsible for only a single sector, need to consider broader influences across sectors. Some researchers have therefore proposed the approach of the water-energy-food (WEF) nexus.

The nexus approach, through enhanced dialogues and collaborations across sectors, considers co-benefits to establish appropriate safeguards of all three resources. [The upper four paragraphs are *the gain-framed info section* and will be called so in the following versions to save space from repeating the exact same texts].

Water, energy, and food security is in a nexus to achieve!

3. Attitudinal Version (One-Sided, Closed-Ended, Gain-Framed)

The Nexus Approach Is a Better Concept to Achieve the Water, Energy, and Food Security!

[The gain-framed info section]

Leading international organizations, such as the United Nations, the International Energy Agency, and the World Economic Forum, have valued the nexus concept as an urgent, important, innovative, and useful approach to achieving water, energy, and food security, a serious task worldwide. [This paragraph is the *gain-framed attitudinal paragraph* and will be called so in the following versions to save space from repeating the exact same texts].

The nexus approach is a better concept to achieve water, energy, and food security!

4. Behavioral Version (One-Sided, Closed-Ended, Gain-Framed)

Let's Promote the Nexus Approach to Achieve the Water, Energy, and Food Security!

[The gain-framed info section]

[The gain-framed attitudinal paragraph]

Organizations and researchers who support the WEF-nexus approach call on everyone to take action to promote the nexus concept to achieve water, energy, and food security. Such actions include donating to support the WEF-nexus research and promotion, telling others about the WEF-nexus concept, and urging others to support and promote the concept too. [This paragraph is *the gain-framed behavioral paragraph* and will be called so in the following versions to save space from repeating the exact same texts].

Let's promote the nexus approach to achieve water, energy, and food security!

5. Behavioral Version (One-Sided, Open-Ended, Gain-Framed)

Should We Promote the Nexus Approach to Achieve the Water, Energy, and Food Security?

[The gain-framed info section]

[The gain-framed attitudinal paragraph]

[The gain-framed behavioral paragraph]

Should we promote the nexus approach to achieve water, energy, and food security?

6. Behavioral Version (Two-Sided, Closed-Ended, Gain-Framed)

Despite the Criticism, Let's Promote the Nexus Approach to Achieve the Water, Energy, and Food Security!

[The gain-framed info section]

Some researchers, nevertheless, criticize the approach of WEF Nexus. They call the nexus approach "the same old wine in new bottles," adding little value to such existing coordinating approaches as Integrated Water Resource Management (IWRM), which balances water management with the management of other resources to meet human needs. Supporters of the WEF nexus, on the other side, argue that the IWRM approach only explicitly focuses on water, and may prioritize water-related goals or interests. The nexus approach achieves water, energy, and food security simultaneously by recognizing their dependence on each other. [This paragraph is *the gain-framed negative info paragraph* and will be called so in the following message to save space from repeating the exact same texts].

[The gain-framed attitudinal paragraph]

[The gain-framed behavioral paragraph]

Despite the criticism, let's promote the nexus approach to achieve water, energy, and food security!

7. Behavioral Version (Two-Sided, Open-Ended, Gain-Framed)

Given the Criticism, Should We Promote the Nexus Approach to Achieve the Water, Energy, and Food Security?

[The gain-framed info section]

[The gain-framed negative info paragraph]

[The gain-framed attitudinal paragraph]

[The gain-framed behavioral paragraph]

Given the criticism, should we promote the nexus approach to achieve water, energy, and food security?

8. Behavioral Version (One-Sided, Closed-Ended, Loss-Framed)

Let's Promote the Nexus Approach to Avoid the Water, Energy, and Food Crises!

Water, energy, and food are essentially linked. Water is needed to produce the crops and cook the foods. Agriculture is currently the largest user of water, consuming 70% of global water withdrawal. Energy is required to produce and distribute water and food. Just the food production and supply chain accounts for about 30% of global energy consumption. Finally, the generation of energy consumes a large share of water withdrawal.

The linkage among water, energy, and food creates interactions and trade-offs. Watering crops to promote food production reduces river flows and hydropower potential. Growing bioenergy crops, such as corn for fuel ethanol, demands more water withdrawal and causes food shortage. To make the watering system more efficient, such as using drip, center pivot, or rain guns, saves water but consumes more energy. The global community is well aware of water, energy, and food challenges but has so far addressed them separately, with fragmented responsibilities, lack of coordination, and inconsistencies among laws and regulations.

However, balancing those trade-offs is central to jointly avoiding water, energy, and food crises at the same time. To do so, decision-makers, even those responsible for only a single sector, need to consider broader influences across sectors. Some researchers have therefore proposed the approach of the water-energy-food (WEF) nexus. The nexus approach, through enhanced dialogues and collaborations across sectors, considers trade-offs to avoid the shortage of all three resources. [The upper three paragraphs are *the loss-framed info section*, with differences from *the gain-framed info section* in the last paragraph, and will be called so in the following message to save space from repeating the exact same texts].

Leading international organizations, such as the United Nations, the International Energy Agency, and the World Economic Forum, have valued the nexus concept as an urgent, important, innovative, and useful approach to avoiding water, energy, and food crises, a serious challenge worldwide. [This paragraph is *the loss-framed attitudinal paragraph* and will be called so in the following message to save space from repeating the exact same texts].

Organizations and researchers who support the WEF-nexus approach call on everyone to take action to promote the nexus concept to avoid water, energy, and food crises. Such actions include donating to support the WEF-nexus research and promotion, telling others about the WEF-nexus concept, and urging others to support and promote the concept too. [This paragraph is *the loss-framed behavioral paragraph* and will be called so in the following message to save space from repeating the exact same texts].

Let's promote the nexus approach to avoid the water, energy, and food crises!

9. Behavioral Version (One-Sided, Open-Ended, Loss-Framed)

Should We Promote the Nexus Approach to Avoid the Water, Energy, and Food Crises?

[The loss-framed info section]

[The loss-framed attitudinal paragraph]

[The loss-framed behavioral paragraph]

Should we promote the nexus approach to avoid the water, energy, and food crises?

10. Behavioral Version (Two-Sided, Closed-Ended, Loss-Framed)

Despite the Criticism, Let's Promote the Nexus Approach to Avoid the Water, Energy, and Food Crises!

[The loss-framed info section]

Some researchers, nevertheless, criticize the approach of WEF Nexus. They call the nexus approach "the same old wine in new bottles," adding little value to such existing coordinating approaches as Integrated Water Resource Management (IWRM), which balances water management with the management of other resources to meet human needs. Supporters of the WEF nexus, on the other side, argue that the IWRM approach only explicitly focuses on water, and may prioritize water-related goals or interests. The nexus approach avoids water, energy, and food crises simultaneously by recognizing their dependence on each other. [This paragraph is *the loss-framed negative info paragraph* and will be called so in the following message to save space from repeating the exact same texts].

[The loss-framed attitudinal paragraph]

[The loss-framed behavioral paragraph]

Despite the criticism, let's promote the nexus approach to avoid the water, energy, and food crises!

11. Behavioral Version (Two-Sided, Open-Ended, Loss-Framed)

Given the Criticism, Should We Promote the Nexus Approach to Avoid the Water, Energy, and Food Crises?

[The loss-framed info section]

[The loss-framed negative info paragraph]

[The loss-framed attitudinal paragraph]

[The loss-framed behavioral paragraph]

Given the criticism, should we promote the nexus approach to avoid the water, energy, and food crises?

Index

Q. J. Yao, *Marketing the Water-Energy-Food (WEF) Nexus*,
https://doi.org/10.1007/978-3-031-93148-2

K

L

M

The manufacturer's authorised representative in the EU is Springer Nature Customer Service Centre GmbH, Europaplatz 3, 69115 Heidelberg, Germany. If you have any concerns regarding our products, please contact ProductSafety@springernature.com

Printed and bound by CPI Group (UK) Ltd, Croydon, CR0 4YY
30/09/2026
02225583-0001